# Living Where the Cats Eat the Dogs

## The Rogue Biologist

Written by
Dan Dourson
2021

*Living Where the Cats Eat the Dogs, The Rogue Biologist* is the first of a two-part series. The second in the series will be *Living Where the Cats Eat the Dogs, The Weeble* (written by my wife, Judy Dourson). Reproduction of any section is prohibited without verbal or written permission from the author.


ISBN-978-0-999-8023-8-0
Published by Goatslug Publications
1128 Gritter Ridge Road, Stanton, Kentucky, USA

Goatslug Publications

**Front Cover:** The beautiful Jaguar art, done by Cheyenne Gordon, a former student of my wife's, is a mixture of different mediums including watercolor pencils, acrylic paint, graphite pencils, and black/white ink.

**Photo/Illustration Credits:**
**pg. 10:** Cockscomb ©*Greg Harris* ~ **pg. 14:** jaguar ©*Ana Vasileva* ~ **pg. 96:** snails ©*Jaap Vermeulen* ~ **pg. 38, 159:** snail ©*Dan Dourson* ~ **pg. 45:** kissing bug ©*Dan Dourson* ~ **pg. 61:** Cartoon ©JP Tolliver **pg. 116:** jaguar ©*Carol Farneti Foster* ~ **pg. 124:** howler monkey ©Nora Prindle ~ **pg. 135:** snake ©*Dan Dourson* ~ **pg. 137:** venom graphic©*Rachael Grabowski:* **pg. 140:** glassfrog © Dan Dourson ~ **pg. 146:** dung beetle ©*Chris Shields* **pg. 149:** peanut headed bug ©Dan Dourson ~ **pg. 161:** cartoon ©*AB Osborne.*

The following images were downloaded with license from Dreamstime.com:
**pg. 21:** snake ©*Anna Kas;* Snail ©*Aleksander Grozdanovski* ~ **pg. 25:** bug spray ©*Jrcasas* #195526625 ~ **pg. 44** Mayan ruin ©*Marcello Celli* #201927601 ~ **pg. 45:** botfly larva ©*Philip Sansonovski* #93457681; scorpion ©*Matthijs Kuiijpers* #24985393~ **pg. 48:** church ©*Viktoriia Protsak* # 12330840 ~ **pg. 50:** rat eating ©*Alexandra Petrak* #39245118 ~ **pg. 54:** butterfly ©*Christos Georghiou* #17087219 ~ **pg. 56:** stick man ©*Zdenek Sasek* #145937360 ~ **pg. 66:** Preacher ©*Eric Basir* #21643323 ~ **pg. 69 & 129:** bat ©*Rudmer Zwerver* #163570144 ~ **pg. 72:** tapir ©*Jeneses Imre* #142453014 ~ **pg. 75:** collared peccary ©Markusmayer1 ~ **pg. 78:** tamandua ©*Isselee* #9332622 ~ **pg. 80:** gibnut ©*Matthijs Kuiijpers* #197715590 ~ **pg. 83:** grub ©*Pisut Chuanyoo* #196921642 ~ **pg. 88:** The Lords of Xibalba ©Diego Rivera ~ **pg. 118:** tardigrade ©*Anna Lopafina* #161887968 ~ **pg. 119:** bonobos ©Alan Dixson ~ **pg. 125:** ~ ocelot ©*Martin Cvetkov* #82493497 ~ **pg. 126:** jaguarundi ©*Patrick Guenette* #163224103 ~**pg. 128:** armadilllo ©*Valentyna Chukhlyebova* #40348380 ~ **pg. 129:** kinkajou ©*Andrii Oliinyk* # 153787152 ~ **pg. 131:** mouse opossum ©*Alexey Malkov* #211377935 ~ **pg. 133:** harpy eagle ©*SaveJungle* #87902251 ~ **pg. 139:** red-eyed treefrog ©*James Kelley* #137080332 ~ **pg. 145:** beetle ©1eveiya1 #157621117 ~ **pg. 148:** praying mantis ©*Bykov Egeny* #92737626 & walking stick ©Vetra Kori #155709771~ **pg. 152:** dragonfly ©*Channarong Pherngjanda* #118610116 ~**pg. 154:** velvet ant ©*Patrick Guenette* #163241482 ~ **pg. 156:** spider ©*SS1001* #78981396 ~ **pg. 170:** man and woman running ©*Igor Sapozjkov* #190959556 ~ **pg. 229:** evolution ©*Tiziano Cremonini* # 120844051 ~ **pg. 128:** silhouette of people ©*Edvard Molnan* #102508393

# Table of Contents

# Foreword

When I was first asked to write a forward for this book, I hesitated, stating "I don't think it really needs a foreword." But after a long walk in the woods (you know that forest bathing stuff), I reached a different conclusion.

You see, I have been tethered (as he would call it) to the man who wrote this book for 27 years. I probably know him better than anyone and as such, should be given my time and thoughts to help introduce him and this autobiography to the rest of you.

The Rogue Biologist, Dan Dourson, is a man of many talents—a Renaissance man in the truest sense of the words. An artist, an author, an inventor, a landscape architect, a carpenter, a gardener, a musician, a naturalist, a paramedic, a firefighter, a machinist, biologist, and most important of all, a father, husband, and friend.

This book started as a medium to tell some of the incredible adventures we shared in Belize, *Living Where the Cats Eat the Dogs*. And while those stories are funny, riveting, and sometimes downright scary, this book is so much more than just a collection of adventures in the rainforest of a zany guy.

Though our original plan was to write this book together, it became clear that our life experiences, while shared on many levels for 27 years, were also quite different with each deserving its own platform, instead of a shared one. So, this book is the first of a two-part series: **Living Where the Cats Eat the Dogs: The Rogue Biologist** is strictly written from Dan's point of view. The second book, **Living Where the Cats Eat the Dogs: The Weeble,** will be from my own perspective.

Writing this book (the first non scientific tome he has written) took Dan on a journey of self that even surprised him. Not many men (or women for that matter) are willing to expose their soft underbellies and dig deep into their psyche to understand themselves more deeply but that's just exactly what he does in this book.

Dan reveals the pain and angst of several undiagnosed learning disabilities that left him feeling dumb. Self-deprecating stories of feeling ostracized for thinking differently, being curious and questioning ANY authority or concepts that were not logical, will resonate with those who have never really "fit in".

He is a man of the highest integrity who is often very painfully honest. Like the time he told the CEO of "Wild Birds Unlimited" that their bat houses didn't work and why. The guy turned and walked away without comment. Dan is often misunderstood for his blunt responses. I learned the hard way that if I really don't want to know his honest opinion about anything—like "does this dress make me look fat??"— I don't ask. While this took some getting used to, I have to say after several failed relationships that were less than honest on some levels, his honesty, however brutal at times, has been an invaluable tool to help me heal.

He also shares with anyone who dares to read the section, his views on religion, a subject that many avoid for fear of offending. But I can assure you, that is not his intention for he seeks the truth, no matter where it leads or how painful it is. He has written this book with the purest of hearts, to share the pain he suffered at the hands of religion and provide commentary that few may have even considered. I hope you find this book entertaining, informative, and most of all, you are a bit more empathetic of those who are outside the box—the Rogues of the world. ~*Judy Dourson.*


**Acknowledgments**

I would like to thank the following organizations and individuals for their valued support in the creation of this book: *Belize Foundation for Research and Environmental Education (BFREE)* for serving as a premier research and education facility and *The Marlin Family,* especially Jacob Marlin, founder and dear friend whose commitment to conservation and the Bladen Nature Reserve (BNR) resulted in the development of BFREE which afforded me the opportunity to study the amazing biodiversity of this area.

I would like to sincerely acknowledge Guy McPherson for firming up my long-held convictions that we have pushed the planet to a place of no return. Guy gave my positions on this most contentious subjects legitimacy, and a recognition that I was not alone in my unconventional way of thinking.


**Dedication**

I dedicate this book to my loving parents, Patricia and Leonard Dourson, who both stood by me through all my many faults and failures as well as my successes. I was no doubt a difficult son who had his own mind from day one and was rarely afraid to speak it. Yet, it was my honor and privilege to be raised by such devoted individuals. I am without a doubt a better person and owe them for my many values and achievements. I will hold dear and cherish the many memories we shared until my day of passing. And last but not least, I dedicate this book to Earth for providing me a place to live and the consciousness to wholly appreciate the gift that it is!


**Reviewers**

A sincere thanks to Mike Dourson (toxicologist) for reviewing the pesticide section, Guy McPherson for comments on the environmental section, and my wife, Judy for her many constructive criticisms which made the book far better. A special thanks to my wife and good friend Johnny Faulkner who's honest criticism, help me rethink the order of chapter's.

# Ah Belize

*The "STAGE" was an endless sea of jade covered mountains,*
*the "PLAY" was simply happiness.*

For most people, travel to another country is like a trip to a Hibachi Restaurant. Little more than spectators, tourists come and go with their rubber-stamped adventures, never really experiencing the "behind-the-scenes" culture. Herded in and out like sheep, their escapades largely mollycoddled like infants. Sadly, most depart never knowing what just lay around the next corner. A new wave of tourism is beginning to take hold in countries like Belize where more value is placed on experiences rather than all-inclusive resorts. Cultural tours that focus on food, chocolate, and experiences such as drumming are emerging.

For Judy and I, our excursions to other countries, especially Belize, have been as participants—living behind the scenes, getting involved and fully immersing ourselves in the place. The deeper connections we made as a result of experiential travel have added a richness to the tapestry of our life experiences, forever changing who we are. There aren't words to describe its significance. While in Belize, I was humbled by its people and enlightened by its vast wonders, from misty mountains to deep blue oceans. It was a place of harmony, where the color of your skin no more mattered than the color of your eyes. It wasn't the lifestyle of the rich and unhappy, it was the lifestyle of the meek and contented. Where happiness wasn't measured in the number of dollars you made, but by the number of friendships you developed. Lively and cheerful was the Belizean theme. Belize was, and always will be, my second home. Need I say more?

John Muir once wrote, *"Most people are on the world, not in it — have no conscious sympathy or relationship to anything about them — undiffused, separate, and rigidly alone like marbles of polished stone, touching but separate."*

# The Rogue Biologist

***Following my Instinctive Nature***

*Once outside the fetters of ordinary thinking, liberty of thought, and imagination are set free to wander and discover new things, or in my case, break free from the shackles of unreasonable social rules.*

From the moment I took my first breath, I have been a rogue with a disregard for normalcy. An early entry in my baby book by my mom confirms that: *Son #2 (that's me) is vastly different from Son #1...(my brother Mike)! He marches to the beat of his own drum.* And I have been different ever since.

How I became a Rogue Biologist in the first place is up for much discussion but Dyslexia and Asperger's probably had everything to do with my "buck-the-system" behavior. While most people follow the path of least resistance, I seem destined to follow the road of most controversy. Like the time I was told by Belizean Park Rangers not to run on the road to BFREE or I would be arrested; the same road we drove to and from BFREE which so happened to cross the Bladen Nature Reserve. Now I am one to follow the rules when they make sense, but when they don't, I often ignore them. So what did I do in response to their silly request? I not only ran the road every morning, but ended by jaunt at the ranger station to make arresting me more easy. After all, I also have a benevolent side as well. As it turned out, they said and did nothing, staring in disbelief at this defiant gringo!

On another rouge-ish occasion, I was hiking in the Corcovado National Park on the Osa Peninsula in Costa Rica with my brother Mike, nephew Aaron, and friend John Hemmerle. When we got to the ranger station outpost, we were all told we had to camp in the shelter provided. All conceded but one. Having no room left for my jungle hammock to hang, I asked the ranger if I could set up my tent in the nearby jungle. He told me that it was too dangerous, a result of some dumbass American grabbing a venomous snake, only to be bitten. Even as I explained to him that more people are bitten by snakes during daylight hours and that people are typically safer and more focused at night using a headlamp, he responded by saying "*a rule is a rule*". I decided this was another one of those ill-conceived regulations, so off to the jungle I snuck to set up my unlawful camp. Two hours later, an officer of the law showed up wielding an AK-47 and clicking handcuffs. I was arrested and taken back to the ranger station where I was interrogated for an hour or so. After a background check, they determined that I was just another clueless American so they released me where I joined my hiking partners all fast asleep.

Ironically, we were told by those same rangers that we had to get up at 4:30 AM the following morning to start hiking in order to reach and cross the river before high tide to avoid being eaten by bull sharks! A few days later, we are on a bus from hell—with a driver on drugs playing chicken with a another truck—above a 500 foot cliff! I guess if venomous snakes don't kill you in Costa Rica, a crazy bus driver just might.

Another occasion would leave Judy and I dumbfounded in the middle of the desert on an Indian Reservation Campground in northern Arizona. Fenced in were the campers with signs saying "*Please stay within campground boundaries*", the beautiful desert beyond the fence is our sacred place. I thought, "That's cool" —that is until I went up to the station and explained to them that I was a biologist and would love to walk among this most amazing place. Their response would leave me speechless. "You can't walk across our desert but we will rent you an ATV and you can drive over it," they replied. I responded by saying, "No Thanks." Later that night, like the rogue I am, I climbed over the fence to enjoy the desert in all its glory, and best of all—leaving no trace!

**Rogue Musings and Mutterings**

When I look at the world, I see two basic kinds of people, those that choose to remain safe and sound inside their boxes of comfort (the majority) and the types that can never find a comfortable seat in those same boxes (me). If it's true that nature never draws a line without first smudging it a bit, then I must be that "smudge" that wandered far from the line of tradition into the world of imagination and uncertainty; questioning the norms of society when they did not make sense to my logical, analytical brain.

Perhaps it all started before I was born, an irregularity in my mother's womb, having absolutely nothing to do with my parents or environment. After all, I have 5 siblings, all raised by the same mom and dad, and yet, they are nothing like me. Snails and whales have more in common than I have with my brothers and sisters. They say that the apple doesn't fall far from the tree, but when this apple hit the ground it never stopped rolling.

As time went on, I would find my way of thinking more and more isolating. My offbeat behavior would only foster a few close friendships, while pushing, well actually repelling, most others away. My dad and I would have endless heated arguments over God and church and, for many years barely spoke to each other. He would disapprove of my lack of religious convictions until his dying breath. But I held him not accountable, after all, this was who he was and I love him for caring so deeply.

I didn't drink or smoke in my youth, nor do I use these social fixations today, not for any religious or philosophical reasons, but because back then my first tastes of alcohol and cigarettes were frankly disgusting. I can't imagine it was all that different for anyone else. My tongue and lungs were saying "this is not enjoyable" and I listened, while most of my classmates were being sucked into the trappings of peer pressure. I have no problem with this stuff. I believe to each their own as long as it's not harming someone else. My point is that the "use of" and "addictions to" these substances probably began not as a result of tasting so damn good, but from peer pressure to conform with the norm, whatever that means. My refusal to use these substances only further isolated me socially.

Even today, at age 67, friends and family continue to try and get me to drink alcohol, or believe in their Gods—claiming that—if I just follow these communal obsessions, I will have a better time and more meaningful life. My response is usually "*That will be the day I need a chemically-induced high to have a good time or believe in fables to have a consequential life.*" Others insist that if I do magic mushrooms or smoke pot, I'll experience nature more fully, be more focused, and more creative! Hell, if I was any more focused, my head would explode! If you're convinced that licking toads or doing ayahuasca will give deeper meaning to your life (whatever that means), more power to you—to each their own. But I remain unconvinced that you're experiencing something more profound in a chemically-altered state than I do sober. I could claim that Moss-Sniffing (which I routinely do) guides my every life decision. While I might view it as a legitimate source of enlightenment, there is no real basis for the claim. I do, however, recognize the growing body of evidence that these naturally occurring substances provide alternative healing methods for the mind and body that traditional medicine does not.

Nevertheless, in my early twenties, I decided I needed a change from the mundane life in Cincinnati, so in a bold decision, I moved to Red River Gorge to live on a piece of land that my brother Mike and I had bought years earlier. It was in the middle of nowhere, near the sleepy little town of Stanton, originally called Beaver Pond. I had left my friends (a total of 2) and family behind and moved to rural eastern Kentucky, where everyone I met was a complete stranger.

I went from living in a typical American home with all the modern conveniences to an 8X10 foot log cabin on a ridgetop with no running water, power, or road. I ditched a comfortable life for an impoverished one overnight, yet strangely enough, never looked back with regret. Of course, my entire family thought I was totally insane for making such a radical move. Even to this day, they don't really understand the allure of my chosen lifestyle. Looking back at it all, I don't remember the unknown as being scary, but the unknown as being more of an adventure. The two years spent in this special place would be a pivotal point in my life, giving me an opportunity to reflect on who I was and where was I going.

Today, while most Americans dwell in typical oversized homes, Judy and I live in a tiny house, off-grid with a composting outhouse. Not because we can't afford a more traditional American lifestyle, but because we actually prefer a simpler existence not bursting at the seams with shit we don't need. It all just boils down to a state of mind. Our basic needs (shelter and food) are met with a few wants (more than one fork and one spoon—all I had when I met Judy). Why just one fork and one spoon? That's simple—less to wash. But having merely one fork and one spoon meant I could only have one guest at a time, and that guest had to choose between one or the other. The downside? I always ended up with the spoon!

Living this way, without any debt, allowed me to quit my traditional job with the US Forest Circus, oops a Floydian slip, I mean a Freudian slip, damn that dyslexia shit! Let me restate it over, I meant to say Forest Service at age 47 (another terrible rogue decision, according to my dad) and Judy at 50, giving us both the freedom to play and travel while most people are donkey-working up into their 70s.

While the unpleasantries of dyslexia and Asperger's were painfully apparent in my youth, for the first time in my life, I now understand why I think the way I do. As a result, I no longer consider these syndromes as brain disorders—but as brain upgrades—a kind of life preserver in an ever-rising ocean of uncertainty. It has made my time on Earth better in ways I probably can't even begin to imagine, and probably why my brain seizes new information like a mantis seizes prey. When I hear a fresh idea based on sound evidence, I am quick to reevaluate and fine-tune my way of thinking according to the newly-discovered insight. But for others, the gears turn more slowly or not at all, especially if it disagrees with their current viewpoint that has hardened over time. *When presented with new information, most people play the game "hide and* ***not*** *seek."*

Thirty-five years ago, I threw my family a lifeline, detailing the science of an unhealthy diet. They were dismissive as they so often were/are with new information from yours truly. In contrast, when I first heard the message, I was all ears and angry for not being better informed on the subject. I immediately got the book "Fighting the Food Giants" and others like it that confirmed what I had just read—*you are what you eat*. So, I changed my diet almost overnight and have been eating healthier ever since. Ironically, most of my family now eats a better diet, although 35 years later, and notwithstanding a number of health problems they have incurred, some likely as a result of a poor diet. As I plowed ever deeper into the human psyche, some other interesting concepts began to surface.

I found that intelligence is traditionally viewed as the ability to think and learn. But according to Adam Grant who wrote "Think Again", there's another set of cognitive skills that might matter more; the ability to <u>rethink</u> and <u>unlearn</u>, especially in this increasingly fast-paced and changing world. Grant suggested it is our cognitive laziness that keeps us pinned down to an idea, preferring our old views to new ones that might require a rethink or an unlearn—religion being one of the best examples. Questioning ourselves makes the world more unpredictable and requires us to admit that the facts may have changed; that what was once right may now be wrong. According to Grant, not everything in our lives is a rethink, such as a new car for example, but when it comes to knowledge and opinions, our gears turn more slowly, if at all. He wrote, "We favor the comfort of conviction over the discomfort of doubt, letting our beliefs become brittle long before our bones."

So, was it a persona set forth by an anomaly in my genetic code that set in motion my Dyslexia and Asperger's. Or was it a conscious or unconscious decision to follow my **instinctive nature** and **passions** instead of humanity's **insanity** and **indifferences**? Perhaps it was a little of both. For whatever the reasons, my oddball personality was the instinctive nature I followed, beginning with my first breaths of life and no doubt lasting until I die. *Hence this is my muddled path to becoming the "Rogue Biologist".*

The rest of the book takes you on my journey through lost mountain valleys, undomesticated river gorges, and deep cave systems, revealing the Mountains of the Maya in all its grandeur as seen through the eyes of an eccentric biologist from Kentucky. Part stories, part science, part philosophy, and every part crazy, it is my hope the following anecdotes will entertain as well as inform the reader of my offbeat perspective while

*Living where the cats eat the dogs.*

Greg Harris

The Jaguar Reserve, Cockscomb Range, Belize

# An Introduction to Belize

From mountaintops shrouded in mist to sapphire-blue seas, Belize is a landscape of varied expressions. The Maya Mountains encompass a low range of karst hills with hundreds of cave systems to mountain ranges with elevations just under 4000 feet. Most of the peaks of the Maya Mountains are more rounded but the spectacular Cockscomb spine is all that remains of the earlier mountains. Scant elfin forests are found on the highest peaks with lowland tropical broadleaf forest at the lower elevations. Annual rainfall in Belize ranges from 70-200 inches.

With a landmass about the size of the state of Massachusetts, Belize's population stands at nearly 400,000 people. Forty percent of that land is under some type of protective status as reserves, parks, or preserves. Its commitment to the conservation of natural areas and biodiversity is undeniable. These protected areas are critical biological reservoirs that restock forests, rivers, and oceans that have been depleted by excess hunting and fishing. Additionally, with ecotourism as its primary source of revenue, it is essential that the environment is protected.

## The Wildlife of Belize

For its small size, Belize has remarkable biodiversity. New species are still being discovered, including the recently described Yucatan brown brocket deer (*Mazama pandora*) endemic to the Yucatan and northern Belize. The longest continuous barrier reef in the Western Hemisphere stretches for more than 100 miles along the Belizean coastline and contains hundreds of cayes and atolls. This astonishing reef harbors more than 500 species of marine fishes, including the world's largest, the 50-foot long whale shark, 106 coral species, and 25 species of marine mammals. The world's largest remaining populations of the endangered West Indian manatee are found in Belizean waters.

With nearly 600 species of birds, 140 land mammals, 150 reptiles, 65 amphibians, 150 gastropods, and countless arthropods, the terrestrial environments of Belize are teeming with wildlife. Ventures into the natural areas promise to deliver a plethora of animal experiences. Birds include the glamorous macaws, parrots, and toucans but also the largest populations in the world of the spectacular king vulture. The Maya Mountains represent the northernmost range of the harpy eagle. While birds are of interest to many, mammals such as margays, monkeys, and the rare cacomistle have their own allure. More than 75 species of bats reside in the deep jungles, caves, and savannas of Belize's wild places.

Amphibians are well-known members of tropical rainforests and one species, the red-eyed treefrog, a stunning example of color evolution, is one of the most photographed frogs on Earth. Bird songs dominate the day but in the rainy season frogs and toads rule the night sounds. Snakes, although abundant, are seldom seen. Around 60 species occur in all habitats found on mainland Belize and most of the larger islands. Only eight species contain venoms strong enough be of concern to humans. Contrary to public opinion, none are aggressive unless molested.

## Life in Belize

A polyglot of cultures coexist in relative harmony with three indigenous groups of Maya (Yucatec, Mopan, and Kek'chi), Mestizos, Kriol, Garifuna, the Mennonites, the Taiwanese, and a whole host of expatriates and retirees from the USA, Canada, and Britain. The pace of life in Belize remains unswervingly simple and largely unhurried, with most Belizeans living close to the land, dependent upon it for their very survival. The simpler way of life is not only conducive to a happier populous but also significantly reduces the threats and negative impacts to the environment and biodiversity of the region. It certainly was a great fit for this Rogue Biologist and his wife!

## Agriculture of Belize

Belize's tropical climate and variations in rainfall are conducive to the production of sugar cane, cashew, aquaculture, citrus, and bananas. Citrus and banana production requires a substantial amount of herbicides, fungicides, and chemical fertilizers to produce high yields with blemish-free products for export. This type of agriculture has proved to be challenging for Belize and its protected areas, including the Bladen Nature

Reserve which sits adjacent to large-scale banana and citrus farms. The impact from chemical drift on the Maya Mountains remains largely unstudied.

Historically, slash-and-burn agriculture has been common practice with devastating impacts on forest ecosystems. Today, new and sustainable ways of producing food while protecting biodiversity in Belize include; crop rotation to boost soils with nitrogen-fixing plants such as macuna beans, adding the revolutionary, age-old technology of soil enhancing "**Biochar**" (a product made from charred wood fiber) to existing nutrient-poor farmland and planting forest-friendly cash crops like cacao, vanilla, black pepper, and coffee.

## A Creative Approach to Conservation

The establishment of protected areas in the 1980s and 1990s was not without controversy. The Cockscomb Basin Wildlife Sanctuary (CBWS) in Stann Creek District near Maya Center displaced families and prevented locals from hunting in an area they depended upon to survive. This set the stage for conflict between residents, the government and conservationists. Locals who were essentially locked out of the CBWS staged a revolt of sorts, cutting trees to block access to the Sanctuary.

Ernesto Saqui, chairman of Maya Center (the village located at the entrance to the CBWS) and the first manager of the Sanctuary, developed a plan that would allow the two entities to co-exist in relative harmony. Enlisting support of the women of Maya Center to create handicrafts made from local materials, Ernesto received permission for ladies to collect entrance fees to the reserve where the hand-crafted items were displayed was obtained from the government, and the Belize Audubon Society (co-managers of the CWBS). Soon the ladies of the community were earning a fair wage allowing them to send their children for further education, and visitors were able to enter the reserve conflict-free. This community-based approach to the problem simultaneously conserved one of the last great strongholds on the planet for jaguars, empowered Maya women through the development and promotion of their art, and set the stage for conservation and ecotourism to coexist. Today, Maya Center thrives with tour guides, taxi drivers, restaurants, and the Women's Cooperative that still sells the ladies' crafts.

A creative solution to the discourse led to an innovative blending of ecotourism and conservation that serves as a model for other communities throughout Belize.

**BFREE**

A plethora of cultures and landscapes provided the backdrop for our incredible adventure in the tiny Caribbean country of Belize where the **"Cats Eat the Dogs"**. For seven amazing years, we were utterly immersed in its culture and jaw-dropping flora and fauna. Every day was an escapade into a world full of wonder and discovery in our newfound jungle home, nestled among ancient Ceiba trees and howler monkeys.

Our home was a biological field research station called Belize Foundation for Research and Environmental Education (BFREE) situated on a 1,156-acre private nature preserve. Our backyard was the Bladen Nature Reserve (BNR), Belize's highest protected area—100,000 acres with no roads and no trails. Referred to by Belizeans as the "crown jewel" of parks, access to the BNR is by permit only and is limited to research and environmental education.

Access to BFREE is a six-mile unpaved road (and I use that term loosely but more on that in the next chapter) passing through a small section of the Bladen Nature Reserve. BFREE also borders three other protected areas including Cockscomb Basin Wildlife Sanctuary, Deep River Forest Reserve, and Maya Mountain Forest Reserve, all part of the larger Maya Mountain Massif covering almost 2 million acres of continuous tropical forest. Considered to be one of the largest tracts of pristine tropical forests left in Central America, living at BFREE with access permitted to this biodiversity hotspot would prove to be the perfect playground for this rogue biologist.

While Judy spent most of her time managing BFREE, its staff, coordinating researchers, and running university programs focused primarily on tropical ecology, I spent nearly all my time chasing, catching, and photographing wildlife. It was a wild ride of monumental proportions for us both that we will cherish until we are planted in the ground. Both family and friends thought we had lost our minds to leave the comforts and security of the USA. Our response? "*We are tired of watching the Discovery Channel, we want to go live it!*" And that's exactly what we did for seven mind-blowing years!

***Albert Einstein famously said, "Imagination is everything. It is the preview of life's coming attractions."***

# *Living Where the Cats Eat the Dogs*

©Ana Vasileva

## *A Journey of Discovery*

# The Road of Lost Soles

## *The Adventure Begins*

*In the beginning, there was but a scant trace through the savanna. Then came a road without empathy for the lost souls of its travelers. In the end, this road would lead many to an adventure of a lifetime.*

Without a doubt, our first day at BFREE was one for the books, a bitter-sweet, mostly road-less journey forever etched in our memories. After an uneventful 3-hour plane ride (aside from my shattered nerves) from the US and 5-hour car ride from Belize City, we finally arrived at Gomez's sawmill, the beginning of our jaunt into a world of the unfamiliar. The next 6 miles would be spent riding in the back of a jacked-up Toyota-Tacoma with a grill made of metal teeth, a vehicle amply named Grizzly. This beast of a truck would carry us on our unimaginable voyage to BFREE. What started out as a proper thoroughfare quickly disappeared, however, into an endless sea of mud that only lunatics dared venture. In fact, the "road of lost soles" is in the top 10 roads of madness, burying any traces of one's sanity below heaps of thick sticky goo.

Driving down what was described to us by Jacob Marlin (founder of BFREE) as the road to Research Station was no road at all, but a 6-mile slog through a soggy savanna. A road aptly named 'road of lost soles' not for the lost souls of men, but for the lost soles of footwear torn from the bottom of shoes during the 6-mile hike in. Half swamp, half mud, little of it road, we would soon find ourselves, along with Grizzly, hopelessly stuck in axle-deep muck.

As Jake struggled to hide his shame-faced grins, I thought, "What's so funny?" Just then Jake turned to me with a shovel and said "Start digging boy!" all the while clandestinely thinking *let's see what this 51-year-old man from Kentucky is made of.* To this day, I believe Jake got stuck on purpose! Two hours later and plastered in mud, Grizzly was finally free.

I climbed back into the truck and turned to my wife in her gleaming white shirt and said "What have we gotten ourselves into?" As we continued our trek to BFREE, something appeared out of the corner of my eye. I had only read about such phenomena in books, but nothing would prepare me for what appeared to be the impossible. Was it a miracle? No, but not far off. A "Jesus Christ" lizard racing across the water—a feat that seems to defy the laws of physics!

They say seeing is believing. I now undeniably believed in Jesus........ Christ lizards that is. Within moments, my reason to why we came to Belize was answered. Judy and I were now living the "Discovery Channel"! As for Judy, her motives for the move would be quite different.

For the next 7 years, the road of lost soles would become the road of my lost sanity. This road was my nemesis and endless tormenter of wellbeing, the thing nightmares were made of—at least for me. I would come to hate this unholy road from hell. In the rainy season, the road frequently disappeared under a blanket of murky water, often turning into a slow-moving river of thick reddish, sole-grabbing mud. Getting stuck more often than not was the norm during the endless precipitation as I moved student luggage and provisions back and forth, to and from BFREE.

If I had a dollar for every shovel of gooey mud I dug from beneath Grizzly's axles, I would have used it to fix the fucking road. You have no doubt heard of road-rage in the city, well, this was mud-fury in the tropics. This was not your every day kind of mud either. It was the kind that stuck to your shovel like a blow-fly sticks to rotting meat. You needed a second shovel to dig out the first shovel to dig out Grizzly. This mud is what tore the soles from shoes and turned my beard prematurely gray.

If I was **God**, this is where I would send bad people to suffer an eternity and if I was the **Devil** looking for a summer home in the tropics, this would be the ideal spot of torment. The road of lost souls/soles could then serve to both punish the wicked and torture the damned.

Having little choice at the time, I would continue to suck it up for the team. Why? Because the environmental education at BFREE far outweighed my road trauma-drama. But given a choice, I would have preferred my naked body being rolled in fire ants than to spend one more second on that river of mud masquerading as a thoroughfare. The road's only saving grace would always be its conclusion at the pristine Bladen River. At this instant, the road madness just moments before would melt away to jungle ecstasy.

# Living Where the Cats Eat the Dogs

## *The Adventure Continues*

As Grizzly, Jake, Judy, and I pulled up to the kitchen at BFREE, three dingo-looking dogs leapt with joy at the new human arrivals. There would be new asses to sniff, faces to lick and free massages behind the ears. It was indeed a good day for the dingoes of BFREE. The dogs, Mayan staff, and smells emanating from the kitchen, (the beginning of *a love/hate relationship with rice and beans*), formed the welcoming committee to our improbable home in the jungles of southern Belize.

It was November before the busy field season commenced, marking the arrival of hordes of American students and researchers that come to learn about tropical ecology. With only a handful of workers, the founder, Jacob Marlin, Katelyn (an intern from Canada), the dogs, Peanut, Butter, and Jelly, it was sure to be an adventure of a lifetime. Little did we know!

The dingo dogs of Belize all look pretty much the same except for color variations. A long lean body, sharp snout, an abundance of energy, and are born to bark like there's no tomorrow. *Little did they know how true that would be.*

Most village dogs are viewed by Belizeans as nothing more than cheap burglar alarms, poorly fed, and by no means, mollycoddled. The pampered pets of the US, with their fancy dog foods, sweaters, cozy dog beds, and any number of other paraphernalia, were as foreign to locals as the sight of a Belizeans pissing off a highway bridges is to us.

Having left our beloved dog, Maggie, behind in the states until we knew if she would be safe from the perils of rainforest living, the small pack of dingo dogs would end up becoming a bitter/sweet experience. For what ever reason, I'm an animal magnet and the dogs of BFREE quickly warmed up to me to the point of being annoying. They followed me everywhere, trying in earnest to be noticed, to be touched, to be loved. Still trying to deal with the separation from my companion and best friend, Maggie, (next to Judy of course), I was reluctant to bond with the dogs. But they were relentless and, in moments of weakness, I caught myself scratching their ears. This bonding process did not last long, for less than a month later, upon returning from a weekend getaway to the beach, we learned that Peanut had been taken, but by who? Or what? The same fate that has fallen upon every other domestic pet at BFREE.

Discussion around the dinner table that night began to revolve around theories about the disappearing dog. The canine-napper had not been seen, but tracks nearby strongly suggested a jaguar. Our Mayan staff from nearby villages informed us that when jaguars begin to age and lose steam, they often resort to hunting easier prey such as sheep, goats, and barking dingo dogs.

For Peanut, it all happened one late night under the house where Tom, the facilities manager slept. The dogs usually stayed there when everybody else was gone to the beach. The dingo dogs, by their very nature of endless barking at every sound, were like dinner bells, calling out, "I'm easy prey, come get me".

The two large cat species residing around BFREE included jaguars and mountain lions. The stealthy jaguar (whose name means killing in a single bound) is a stalker, with pads the size of a saucer, allowing this magnificent beast to move through the jungle virtually unheard. Jaguars are the third largest cats on Earth, larger than leopards but smaller than lions or tigers with the strongest bite of any cat, used to crush turtle shells to access the flesh.

No doubt, Peanut's incessant barking attracted the attention of a stealthy jaguar that crept up to the house and with one swift pounce and powerful bite to the back of the dog's neck, ended Peanut's life in mere seconds. Only a quick yelp was heard. Thankfully, Peanut suffered little, for the cat's efficient hunting strategy was, if anything, merciful. And so, we were now down to two dogs, Butter and Jelly.

Days turned into weeks and our dog count was still at two. We were all beginning to breathe a sigh of relief when our first family members arrived for a visit. My brother Mike, a renowned toxicologist and his lawyer wife, Marti were well-traveled. Mike spent a fair amount of time in remote Africa, so was no stranger to the bush and was also a scout leader, while Marti who grew up on a farm in Ohio, had never experienced anything as remote and wild as our biological field station. Neither were strangers to the somewhat primitive pit toilets or even the cold outdoor showers—the least of their worries here.

The rainforest is filled with sights, sounds, and smells that can evoke both reasonable and unreasonable fears. With palm trees that bite and bugs the size of rats, it can be disconcerting place to say the least. But their nerves were clearly rattled at the breakfast table when Kate announced that Butter was missing. After breakfast, we all became Sherlock Holmes and spent the morning investigating the dog's disappearance. Once again, the familiar paw prints of a jaguar were discovered nearby, but no evidence of Butter, no doubt taken by a big cat.

The last dingo and now quite skittish Jelly stuck to everybody like glue, especially Katelyn. Who could blame her after witnessing the disappearance of her siblings to a hungry feline? Jelly lasted almost to the end of the field season but would also succumb to the crushing jaws of death. Kate, whose quarters were in the former teacher's house was asked to move in order to accommodate a guest who required a private room and Jelly, of course went with her. First, she put her jungle hammock underneath our raised wooden house where Jelly slept nearby. During the daylight hours while Kate worked in the garden, Jelly would rest on the porch outside our door, oozing drops of blood from her botfly lesions, while I worked on the Belize Biodiversity Book inside. When space at the bunkhouse opened back up, Kate moved over there in order to give us and her more privacy.

A day or so after the pair left, I was abruptly awakened in the middle of the night by the sway of our stilt house. Something or someone was on our staircase, so I grabbed my flashlight and descended our sleeping loft to check out the intruder. It was a freaking jaguar on our porch, attracted to Jelly's dried blood from nights before!!

Although Jelly was spared that night, sadly the stealthy jaguar would eventually catch up with her as she slept under the open area of the bunkhouse. As you can imagine, Kate took the news rather badly, having grown attached to Jelly. As I wrote this account, I realized that the dogs actually disappeared in the order of that great American-style lunch, Peanut.....Butter...and Jelly. And so began our first year of this amazing adventure, *Living Where The Cats Eat The Dogs.*

## Close Encounters of the Best and Worst Kinds

It all happened one late night in the jungles of Belize. We had just finished another endless "Groundhog Day" dinner of rice and beans. Judy left to close-up the chicken coop door to keep the local vampire bats from feeding on their blood, while I remained in the kitchen wolfing down freshly baked monkey-poo cookies (a BFREE original, recipe at end of book). Just another ordinary night in the jungle, right? Wrong! It was at that moment, I heard the sound every respectable biologist longs to hear— "SNAKE". Even as every other levelheaded person was running in the opposite direction, I was jogging toward the blaring voice of my wife with both apprehension and excitement.

Was it a Tommy Goff, the local name for Fer-de-Lance, or any number of more than 50 species of snakes known from the area? I was now in overdrive as I raced toward the glow of Judy's headlamp on the path to the chicken coop.

And there it was in all its magnificence—**A close encounter of the best kind**—the most dangerous and feared snake in the Americas and, according to the local Maya—a bad ass hombre, it indeed was a fer-de-lance! Armed only with this information, I didn't really know what to expect. So caution ruled that night for fear of an antagonistic and deadly encounter. To my surprise, however, the snake was completely docile, showing little signs of hostility toward me, even after a bit of harassing. This was done out of curiosity to see how the snake would react since one of my jobs while at BFREE was to catch and relocate venomous snakes away from buildings where they might encounter people.

During the seven years I spent in Belize, I have quite literally stepped on, walked by, and placed my hands on top of those one-inch long life-taking-fangs. A largely docile serpent in my experience, the aggressive reputation of the Fer-de-lance is unfounded. The venom of the snake, however, is not fable and can kill a human within hours and, like a loaded 357 Magnum, demands the utmost respect.

This no doubt brings up a point of contention in the minds of many: *that all venomous snakes are dreadful creatures, waiting in the grass to attack and bite.* Nothing could be farther from the truth!

**Close Encounter of the Worst Kind**

While venomous snakes deserve maximum respect for obvious reasons, non-venomous snakes can also cause unanticipated problems. Many years ago, while teaching an outdoor class on Kentucky snakes, I had a **close encounter of the worst kind** and nearly died, not from the bite of a venomous serpent, but the bite of a non-venomous northern watersnake. I was holding the snake when it bit me on my right thumb, and in a dumb but habitual response, jerked my hand backward leaving my thumb with deep lacerations from the snake's razor-sharp teeth. I rinsed my bleeding wounds out in a nearby pond, thinking nothing more of the incident, and continued talking. After all, I had been bitten by hundreds of non-venomous snakes in the past with no ill effects.

That night I spent an hour digging out snake teeth from my sore thumb. The bite got infected from what I presumed was bacteria in the snake's mouth or from the pond water. Nevertheless, after a week of soaking my thumb in hot salty water, it began to heal with one unforeseen side effect, a swollen lymph node under my right armpit. No big deal, right? WRONG! The repercussions from the bite weren't over, but just the beginning of something that would put me on the razor's edge of death.

That same week, I headed down to western North Carolina to begin a biological assessment project on land snails for the US Forest Service. While stretching to turn off my camp light that night, I felt what seemed like a slight tear under my right armpit, but no pain. Unbeknownst to me, I had ruptured the lymph node which promptly flooded my system with a toxic concoction of God knows what. Not feeling any ill effects at the time, I went to bed. The next day I would begin to feel the effects of an invasion of the worst kind. Each hour would worsen my condition, so I decided to drive home to Kentucky. But first I had to hike—at times crawl—back to my truck that was now miles away. Finally in my truck with a now useless right arm from swelling, I had to shift gears with my left arm while throwing up several times along the way home. But I managed to reach Richmond, Kentucky where I basically just collapsed. From there, Judy would rush me to the emergency room.

My normally low blood pressure had dropped to just 60 over 35 and I was admitted to ICU where I remained for 5 days, being pumped full of the hardest hitting antibiotics known to medicine. According to the doctors, I had a 13% chance of surviving the ordeal. I remember the infectious disease doctor also told me during his visits that if I survived, I would likely lose my right arm which was now three times its normal size, blistering and peeling like an over-cooked chicken. I was on IV antibiotics for six long weeks. Ironically, my good friend Jim Harrison was milking a king cobra for venom research and was bitten by the snake about the same time, but would be released from the hospital days earlier.

So the moral of this embarrassing story is simple. If you are bitten by a northern watersnake, don't wash your bite out in pond water or you may suffer the same fate or worse. Furthermore, if you handle snakes or any other wild animal, always expect the unexpected.

## Snails Over Snakes

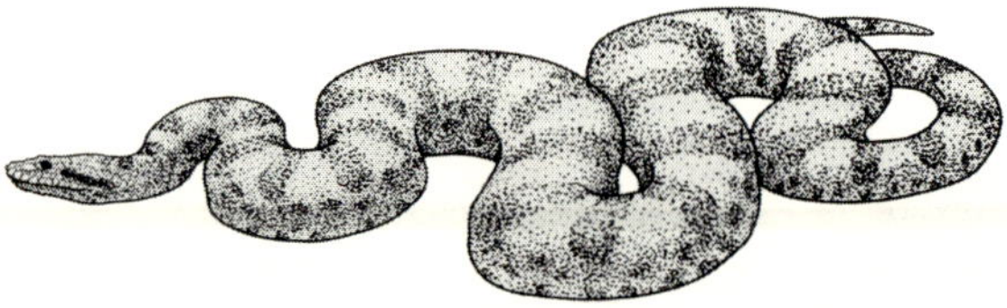

When I got to BFREE, I was immediately interested in learning more about the biodiversity of the Maya Mountains and a little-known group of animals, terrestrial gastropods (snails). Being a neophyte to the rainforest environment, I felt as if I was born again, surrounded by unfamiliar sights, sounds, and smells including colorful and strangely-shaped gastropods. Night walks in tropical rainforests (any forest for that matter) offer truly amazing encounters with life on Earth. The more time I spent in the deep jungles of the Maya Mountains, the more I yearned to understand its secrets. So I was more than ready to crawl into the wonderful world of tropical land snails. Of course, Jake being the smart ass that he is, asked me, "Why in the hell would anyone in their right mind study land snails?"

My response?

> "Well that's just it, Jake, I'm not and never have been in my right mind which makes it easier to step outside the box of reason into the realm of crazy. As it turns out, I am also a well-seasoned smart ass, just ask my wife Judy. Why study land snails you ask?
>
> It is no accident, Jake, that 95% of the "animal diversity" on planet Earth is comprised of invertebrates (arthropods, arachnids, myriapods, mollusks including **SNAILS** and corals, around 1,305,250 described species), the "lesser-majority", while only around 5% are vertebrates (mammals, birds, fish, reptiles including snakes and amphibians, around 66,178 species), the greater-minority. There are roughly 350,000 currently named species of beetles alone, more than 6 times the entire number of vertebrate species combined.
>
> Clearly, along with plants, invertebrates are the driving engines of life. Without their significant and largely uncalculated services, many ecological systems simply cease to function properly, or at all. For many animals, insects are the power bars of the food web, especially important to neotropical birds such as the Kentucky warbler before its winter migration southward to Central America. Many plants would disappear entirely if not for native pollinating bumblebees and hoverflies. With such diversity and adaptability, some say arthropods will inherit the Earth, while entomologists emphatically state that they already own it.
>
> So this is my convoluted environmental journey and the impetus for the study of what most clueless, snake-loving, snail-hating morons are incapable of wrapping their insignificant tiny brains around."

Unaffected by my psychological digs, but mildly impressed by my long -winded answer, Jake replied, "But snakes are so much cooler than snails!"

"Really? Can snakes screw nonstop for days? Snails can. Can snakes have sex with themselves? Snails can. Can snakes shoot calcareous love darts into their mates. Snails can. Clearly, snails are far more interesting than snakes!" I responded.

And so for the next 7 years, the teasing and belittling between Jake, the Snake and Dan, the ham remained unyielding. Jake would continue to poke fun at snails the way Trump makes fun of science. Boy, that felt good to compare Trump the chump with Jake the Snake!

But Jake was relentless and would continue to question the benefits of snails while my scripted one-liner response would eventually become "*snails are the building blocks of the ecosystem*" which Jake would come to hate. But hate eventually turned to undeniable respect for those shelled creatures. And while Jake the snake would never admit that snails are cooler than snakes, he can't deny the fact that land snails, not snakes, put BFREE on the map.

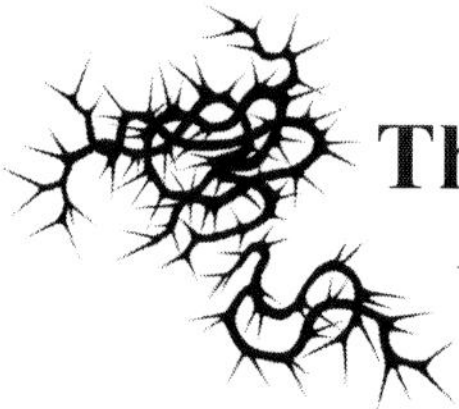

## The Jungle is No Walk in the Park

*Everything out here will stick, sting, or bite you, and that's just the plants!*

What comes to mind when you hear the word *jungle*? Snake-infested, hot and humid, buggy? Danger around every corner?

The jungle is many things but it's not snake-infested the way you think. While there are more kinds of snakes in a tropical jungle than in a temperate forest, there aren't necessarily more snakes. The real "catch" in the jungle are the plants; vegetation that will stick, sting, and at times, bite an unwary intruder showing little mercy. Grabbing the aptly named Give-and-Take Thorny Palm, its spines are plunged deeply into your flesh, *giving* you immense pain but the tree *takes* them back, a result of being well-attached to their trunks.

Then there's the iniquitous Basket Tie-Tie palm or Devil's-Penis as it's called in Belize. The brittle thorns need only touch the skin to break off and lodge. I remember a student grabbing a Devil's Penis one time that delivered hundreds of painful thorns; taking us hours to dig out most of the bodily assaults. I have even been stuck by these He-devil palms just by stepping on one which delivered inch-long thorns into my calf that remain to this day.

There are stinging nettle relatives on steroids that contain stinging hairs from Hell, acacia trees with 5-inch long thorns loaded with fierce ants, and poisonwood trees that can give you a skin-blistering rash just by looking at them. Most of these are defense mechanisms that have evolved to protect ripening fruits and seeds from being prematurely consumed by animals.

Plants in the jungle and their ability to deliver excruciating pain or poison are not restricted to the tropics either. I learned this mouth-burning lesson years earlier in Kentucky. One of my first and most unforgettable encounters with a native wildflower (Jack-in-the-pulpit) was with my good friend Wallace Booth. One day while working in the woods, Wallace convinced me that the corm (root) of the plant was very tasty and that I should try some. Being a neophyte from the city and anxious to try anything new, I didn't hesitate to sample.

As Wallace struggled to hide his hangdog grins, I thought, "What's so funny?" The first few bites were rather bland, but what followed was nothing less than fire in the hole! Needle-like calcium oxalate crystals were flooding my mucus-membranes. Aptly, my opinion of Wallace went from good old boy to you merciless bastard. Unable to speak coherently and give Wallace a proper tongue-lashing, I had suffered my first uninformed encounter with a plant with altitude. The awful sensation lasted almost as long as Wallace's hearty laughter.

I had just received a "bonafide city-slicker initiation"—a common practice here in eastern Kentucky. It never seemed all that funny until my brother John, daughter Angie, wife Judy, and sons Brad and Tyler, all somehow experienced similar mouth-burning encounters. Some say pay it forward. My family and friends say, "Don't eat anything Dan pulls from the ground!"

If you ever have the privilege of visiting a jungle, it would behoove you to pay more attention to those **plants** that stick, sting, or bite, than those **critters** that are mostly crawling, climbing, or running away from you.

Hot and humid describes the lowland jungle very well but higher elevation cloud forests can be downright cold, especially during the rainy season! I have spent colder nights in the rainy Maya Mountains of Belize than in the high snow-covered Sierra Mountains of California.

Buggy you say? Yes, but like snakes—not what you might imagine. For example, there are more kinds of bugs in the jungle than any other place on Earth but most 99.9 percent want nothing to do with *Homo sapiens*. The few groups that do such as mosquitoes are actually in much smaller numbers, and nowhere near the multitudes you see in the Florida Everglades or farther north in Alaska. Finally, danger around every corner is a myth. Judy and I have spent countless nights in the jungles where jaguar and mountain lions walked our same path, swam with crocodiles in Belize, caimans and piranha in Peru, and handled spiders the size of saucers and lived to write about it.

## Is Chemical Warfare Really All That Necessary?

Even though the jungle is full of creepy crawlies, most critters want nothing to do with *Homo sapiens*. While this is obvious to the locals, many visitors believe that everything in the jungle is out to get you. Although most critters will defend themselves if threatened, they are not typically looking for trouble—there's plenty of that in the jungle! For the few arthropods that hunt us for our life fluids, there are a number of methods that can be used in their suppression.

As masses of student groups descended on BFREE, one thing became all too smelly-clear to me, *fear of the blood-sucking, disease-filled mosquitoes and other biting insects that caused them to douse themselves in chemicals*. As the guy who would take them on guided night walks to experience the wonders of the jungle, the smells emanating from their chemical-laden bodies, were at times, overwhelming. I have seen students using 95% DEET applications directly on their skin and report tasting the chemical in their mouth 10 minutes later and melting plastic parts on their binoculars. I can't imagine what it does to a person's central nervous system as it enters their bloodstream through their skin!

In our opening welcome to students, I was in charge of the health and safety talk and I always began with a few tips about how to keep those biting buggers at bay. And none of those tips included the use of DEET. My lecture usually began with this. The best defense against the 1% of biting insects is not necessarily chemical warfare but to simply cover up by wearing loose, long-sleeved shirts, and long pants. I explained to them that most blood-seeking arthropods such as mosquitoes and ticks are found from ground level to around four feet where exposed legs of mammals (humans included) are easily accessed. In addition, mosquitoes are most active during the early hours of the night, so standing around allows them to zero in on your out-gassing carbon dioxide and foul body odors. The more you sweat, the more they come! Moving around instead of standing around during conversations will help foil their ability to find your body signals which are screaming out like a flashing neon sign.

Standing around should also be done with caution, with regards to ants, which if present will soon find your exposed flesh and deliver painful stings and bites. Furthermore, you don't want to be standing next to or on top of a venomous snake, for obvious reasons. Our motto while in the jungle is always "**check the land before you stand**".

But most students insisted on wearing shorts anyway, and by the end of the week looked like they had contracted measles, even though they had applied ample amounts of DEET. If you decide to use a repellant, two types can be applied. First, of the synthetic chemical types, DEET is the most widely used and one of the most effective repellants on the market today. But caution should be used. National Park employees that have used insect repellents containing DEET frequently have complained of chest pain, wheezing, muscle cramping, skin rashes and blisters, dizziness, disorientation, and difficulty concentrating. Other reports of prolonged use of DEET have included adverse neurological effects, seizures, uncoordinated movements, agitation, aggressive behavior, low blood pressure, and skin irritation.

According to research published in *The New England Journal of Medicine,* however, most cases of serious side effects caused by DEET involve long-term, heavy, frequent, or whole-body application of higher percentage of the repellent. For best results and safe levels, 20% to 30% DEET content is recommended. When it's applied only on exposed skin for short periods of time, many researchers believe that DEET can be used safely avoiding many insect-borne diseases. In my opinion, the jury is still out on the safety of DEET. Check out EPA website for additional information on safety and use.

Personally, I would rather ***take a Beaten' than a Deetin'.***

There are the natural alternatives that contain lemongrass, cedar oils, or citronella extracts which shown to be nearly as effective as products containing the controversial DEET although shorter-lived by hours requiring applications more often. Or you can do like the natives and smear sticky mud over your entire body. When the clay dries and hardens, the biting mosquitoes, unable to stick their pathetic beaks through the crusted clay, are unable to feed. You also get the health benefits of a mud-rub at the same time. Judy and I lived in the jungle for 7 years without using a single drop of that smelly and potentially risky DEET just by following those few simple rules.

Nevertheless, we would all prefer ticks and mosquitoes take us off their menu. The primary tick species occurring throughout Belize is the Lone Star tick. The good news is that ticks in Central America are most active during the short dry season, becoming almost absent during a much longer rainy season. The bad news is that ticks in Belize are repotted to carry *Rickettsia rickettsii*, or Rocky Mountain Spotted Fever. And then there are the **"ticking tick bombs"** of the Lone Star Tick.

After the egg, most ticks have 3 stages, called instars, starting with the larva (6 legs), nymphal (8 legs), and finally adult (8 legs). While most tick larvae move around singly, the larvae of Lone Star Ticks hang together in tight clusters of 300 or more. That means if you brush up against one of these ticking tick bombs, they all attach to you at once, leaving you covered with hundreds. What begins as a spot on your pants or exposed legs quickly grows in size as the specks move away from each other, looking for that perfect spot on your skin to stick their soda straws in and begin the siphoning. A little more good news, however, the larvae of ticks, although annoying, have not yet fed, so are without pathogens. That comes later with the next, and final stages of maturing ticks.

Duct tape works best for removing tiny ticks from clothing and skin, so carry it in your pack. Attached ticks can be removed with fine-tipped tweezers. Apply the tip of the tweezers (dipped in alcohol) to the head of the tick (not the body), and remove. Clean the site with disinfectant. And if you really want to get undomesticated, instead of throwing away those chewy little ticks, just pop them in your mouth like the monkeys do. They taste a bit like well-seasoned human earwax and might even give you a natural immunity to some malicious parasite. By the way, did you know that human earwax is nature's perfect insect repellant? It's the reason roaches don't enter your ears at night and lay eggs. Don't believe me? Try tasting it sometime. As nature intended, it is downright repulsive but in a reassuring way.

Even if you opt for chemical warfare, at the end of the day, always check yourself and others, leaving no hair or fold of skin unturned, especially those unmentionables. I have many friends that have used sprays containing DEET or Permethrin, skipping their nightly checks, only to contract Lyme's or some other disease later. Ticks can and will get through even the toughest chemical defenses.

## To Hell with Those Blood-Sucking Parasites—You Say

As a biologist that thinks outside the box of normal reasoning and, in defense of those blood-sucking parasites we hate, I would add these few thoughts. While it is easy to see the value of animals like harpy eagles and jaguars, parasites such as ticks and mosquitoes can be harder for us humans to wrap our self-absorbed brains around. But nature doesn't care if we like or even understand its evolutionary madness. If humans were to disappear tomorrow, nature would continue without missing a beat while the absence of nature would end ***Homo siphons*** overnight.

What's a ***Homo siphon***? *A close, de-evolved relative of Homo heidelbergensis, that, because of its technological advances, absent ethical morals—siphons Earth's natural resources like there's no tomorrow.*

We like to imagine the universe is all about us but modern humans have only been around for the last 300 thousand years or so while the Earth has been in existence for over four and a half billion years, making us almost entirely tardy to the party. Parasites, on the other hand, have been around nearly the entire time, shaping healthy, thriving ecosystems across the globe. The group's overall ability to help flora and fauna reach equilibrium is at best poorly understood.

Research suggests, however, that parasites such as blood-sucking arthropods maintain fitter ecosystems by transmitting secondary parasites to their hosts, which during stressful times, help weed out the weak while favoring the strong, thereby producing a more robust gene pool. Other benefits would include population reductions to more sustainable levels. Furthermore, parasites can provide information on population structure, evolutionary hypotheses, environmental stressors, trophic interactions, biodiversity, and climatic conditions, all of which help us better understand the planet we live on.

## Rethinking Chemical Warfare—MY PET PEEVE

Regrettably, most of humanity has now been programmed into thinking that only one answer exists for pests, "Chemical Warfare". The mindset nowadays is to spray the bugs in your house or the weeds in your yard. I have friends that even spray Roundup on poison ivy, never appreciating the vine for its beauty and the fact that it provides over 60 different species of songbirds a much-needed food source during harsh winter months. It seems rather ludicrous to employ a potentially worse poison to kill a more mild poison. But there are, in fact, non-anthropogenic chemical solutions like simply learning to live with these so-called pests. I have spent my entire life avoiding the use of insecticides and weed-killers around my house and am no worse for it!

One of the best ways to deal with a problem species is to simply give them what they want, away from where you don't want them. Problem solved without the use of chemical warfare! For example, when we get overrun with sugar ants in the kitchen, I simply buy a bag of lollipops, stick a few in the ground outside the house, and within an hour, all the ants are gone from the kitchen. After finding the lollipops, the ants have what they want, a sugary treat, and better yet, didn't have to walk as far to get it. If we get too many roaches in the house, I simply catch a few wolf spiders and release the eight-legged arachnids of wrath upon the roaches. If I have crow or groundhog problems, I simply placed a pile of corn at the end of the garden, giving the marauders what they want!

It always saddens me to watch Americans spraying Roundup on dandelions while never taking the time to fully appreciate this marvelous plant for its beauty. It has delicious, nutrient-rich leaves that can be added to salads or those wonderful fritters made from the flowers, not to mention its food value to bees and other pollinators. And what child doesn't enjoy picking a handful of yellow or blowing those dried seed heads in the wind? I know I did! Hell, I still do!

I remember seeing a little girl weeping after her favorite dandelion flowers went from bright-yellow to wilted-brown, a result of chemicals used by her father to rid the yard of her beloved plants. If it's true that beauty is in the eye of the beholder, perhaps we all need to take a step back from our preconceived notions, *that the only good weed is a dead weed* and look at how our children see a world not fixated on the status quo. Like choosing a weed-less, chemical-laden yard over a lawn that's safe for our kids and pets to play in. My axiom has been and continues to be, "Why spray it when you can eat it or enjoy watching the bees feast on the nectar?" Besides, there are plenty of natural substances out there, some far more toxic than anthropogenic ones. Botulinum toxin, produced by anaerobic bacteria, is one of the most toxic substances known, not to mention Arsenic, Poloium-210, and Mercury. So why add more to the mix?

According to toxicologists, there are around 30,000 human-made chemicals in active marketing today. Many of these are naturally occurring chemicals that are otherwise part of products and "Generally Recognized As Safe" (GRAS). These latter chemicals may not have any testing (e.g., sugar). All other chemicals have some degree of testing, although some of this is minimal (i.e. a structure and lethal dose test). Those which are part of drugs, crops, or food have extensive testing costing a million dollars or more. Numerous websites say that humans are exposed to thousands of chemicals without testing. While this is true, it really only applies to natural chemicals (e.g., coffee has ~1000 chemicals in each cup), few of them tested for toxicity, and they generally fall into GRAS, But not all, 20-30 are known carcinogens.

While anthropogenic compounds have made our lives better on a number of levels, there is no shortage of ever-evolving problems. For example, using a tsunami of chemicals to control the "Green Peach Aphid" has caused the bug to develop resistance to over 70 synthetic chemical insecticides. More concerning though are the declining numbers of beneficial insects (bumblebees, spiders, ladybugs, and praying mantis) caught in the crossfire of pesticide use.

As if anthropogenic pesticides weren't enough, there are natural pesticides made by plants designed to ward off insect attacks. Although they can be quite toxic as well, they usually don't kill the invading bugs, they only repel them. The bugs simply go somewhere else for a less toxic meal, and as a result, strike a kind of crude suggestion of balance. For there is rarely ever a true balance in nature.

This rudimentary system of "**repel and switch**" keeps plants from being over-grazed and prevents the annihilation of any particular species. After all, a robust ecosystem doesn't work as well without all its cogs in place.

***Diversity is key to ecosystem stability*.**

Fifty years ago, a handful of anthropogenic chemicals and predatory insects, bats, and birds took care of what we consider pests. We now have thousands of chemicals on the market, yet have tenfold more pests and they are becoming more resistant to our synthetic pesticides resulting in increasing amounts used and ever stronger poisons. In many areas of the US and across the globe, ecosystems are coming apart at the seams by the overwhelming use of anthropogenic pesticides.

While improper and or over-use of pesticides has resulted in untold fatalities in beneficial insects, some research suggests these chemicals may also be compromising the immune systems of birds, bats, frogs, and salamanders, all of which already have vastly declining numbers worldwide, mostly from habitat loss.

Two recent and concerning studies from Germany and Puerto Rico have concluded that there is an ongoing global "insect apocalypse". Without insects, Earth loses all reptiles and amphibians and about half of insectivorous bird species. The avian seedeaters will eventually fall as well, as a result of declining seed production tied to insect pollinators. And that's just the tip of the iceberg of what is to come as insect populations crash around the globe. Ninety-five percent of Earth's animal biodiversity are invertebrates, not vertebrates that makeup only 5 percent.

According to the United States Department of Agriculture, more than 500 million pounds of chemicals rain down upon US soils every year, killing untold numbers of targeted as well as non-targeted species. This is a huge problem that should concern us all. Why? Because approximately 80 percent of all of Earth's plants are angiosperms or flowering plants that comprise around 50 to 90 percent of human diets and most require insect pollinators to bring that food to our table. It has been estimated that one out of every three bites of food are the result of pollinating insects.

Our fragile environment has already or may be reaching a tipping point of no return and guess who is not becoming more resistant, long-lived animals such as humans and bats to name a few. I fear these "Slaughtercides" are not short-term solutions but long-term disasters with lasting, irreversible effects on Earth's fragile ecosystems. The idea of being part of some large experiment or guinea pig for chemical companies whose employees sit in windowless labs mixing chemical concoctions designed to assassinate life on a large scale is concerning, to say the least.

If biocides were not enough to worry about, the overuse of nitrogen and phosphorus fertilizers on lawns and farmland are causing cyanobacteria blooms (blue-green algae), which in turn causes dead zones in lakes and oceans, threatening biodiversity across the globe. Moreover, there is no easy path to reducing the use of these two fertilizers without creating a bigger risk of food insecurity. A few solutions that have been suggested include the use of Biochar and companion planting of legumes to build back soil integrity.

In Belize, banana farms are biological deserts as a result of their monoculture design but also from a cocktail of toxic chemicals including thiabendazole, imazalil, azoxystrobin, myclobutanil, pyrimethanil, imidacloprid, ethoprop, oxamyl oxime, buprofezin, tebuconazole, and chlorpyrifos used in their production. It is not uncommon to see dead birds and non-target insects scattered throughout a banana farm.

As part of tropical ecology field courses, Judy took groups to these farms to learn how bananas are raised. Although most went in happy and excited for the experience, many left in tears, swearing never to buy another chemical-laden banana again. But for workers on these farms, there is little escape from the ravages of pesticide poisoning.

The good news is that Fyffe's, the British company that buys most of Belize's bananas, was "forced" to relocate many of its immigrant workers from Guatemala, Honduras, and El Salvador out of the interior of the farms due to excessive illnesses caused by being subjected weekly to aerial chemical spraying of these farms. The fact that the Fyffe's was forced to relocate the workers to began with, is in itself, disturbing!

While this has helped reduce pesticide exposure among workers, men who work on these farms are still becoming critically ill with melanoma and penile cancer, while women suffer from leukemia and cervix cancer—all so we can have that perfect, blemish-free banana. We raised non-monoculture bananas at BFREE without the use of pesticides and although they were not without imperfections, they were, in fact, the best bananas to pass through my mouth!

While on the subject of pesticides and the effects on human health and our ecosystems, imagine the following dialogue on food production and pesticide use between a chemical company and the oldest farmers on Earth, the leaf cutter ants and their Queen.

**Chemical company (CC) to the Queen:** *If you use our chemicals, you can grow more food.*

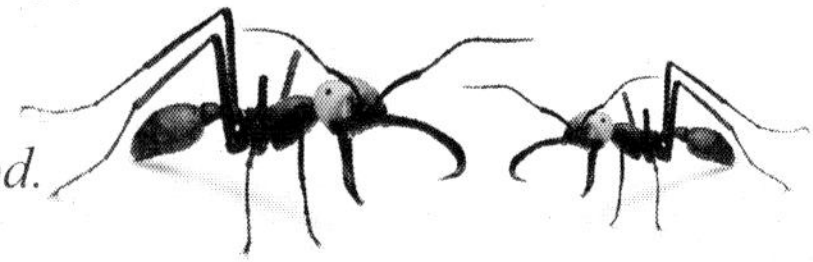

**Queen:** *But we already grow enough food.*

**CC:** *With our GMOs, you can grow bigger food with a longer shelf life.*

**Queen:** *But we prefer smaller tastier food over bigger size and prefer fresher food over longevity. In other words, quality not quantity! Why grow more food than we need anyway or give up taste over appearance? This is silly. Besides, your species throws away half of all food you raise. What a terrible and shameful waste of precious resources!*

**CC:** *Yes, but there's plenty of Earth left to consume and with our super GMOs and pesticides, you will have no pest problems.*

**Queen:** *Really? I understand that your company continues developing newer and more toxic compounds to deal with an ever-growing number of resistant pests? Is this true?*

***CC:*** *Yes, but we have studied these chemicals for a few years now and consider them safe.*

***Queen:*** *A few years doesn't seem like a lot of time to make such a guarantee.*

***CC:*** *Trust us. We have been in the chemical business for 75 years and we know exactly what we are doing. And we pay experts a lot of money to tell us this.*

***Queen:*** *We have been in the business of farming fungi for more than 10 million years. I think we will trust time-proven and clean solutions over short-sighted goals and profit motivations. Besides, WE take care of pests using a formic acid compound. The pests don't develop resistance, and best of all, it doesn't cause birth defects in our young.*

***CC:*** *You are making a big mistake not using our products.*

***Queen:*** *Time-proven evolution a mistake? We ants think not!!*

A contrived dialogue to be sure but this is exactly how we humans think, stuck in a quagmire of indoctrination that our food can only be raised with the use of anthropogenic chemicals. But things are slowly changing to more holistic and sustainable ways of farming. Like treating soils and the plants we consume as living, vibrant ecosystems, instead of lifeless, chemical warzones. The ironic thing is that growing more food and using more pesticides has not saved more people, which are starving in ever greater numbers.

While the "**short term**" effects of chemicals such as glyphosates (Roundup) on human health and ecosystems have been well studied and reported to be of minimal concern, "**long-term**" effects from these chemicals are yet to be demonstrated. These pesticides have not been around long enough to make sound scientific assessments.

Perhaps the more important question we should all be asking is this: "How do the hundreds of chemicals we spray in our houses, yards, and farmlands interact when mixed together?"—a nearly impossible study to conduct—let alone answer. Furthermore, what are the "accumulative effects" of such a concoction of chemicals, coming at us from every angle, including, but not limited to the foods we eat, the water we drink, and the air we breathe? While the lead used in one garden hose may be considered safe (according to EPA standards), at what point does lead from 100 other sources we are being exposed to reach an unsafe dose???

Pesticides that were once considered safe like DDT are now banned in the US but still used elsewhere and, while there have been short-term gains in the control of mosquitoes that carry malaria (a good thing), long-term damage from endocrine-disrupting chemicals, specifically DDT metabolites, have been implicated in reproductive failure in the American Alligator. DDT and DDE have also caused eggshell thinning and embryo death in brown pelicans, Mariana crows, Peregrine falcons, Mariana gray swiftlets, the California condor, wood storks, and marbled murrelets. Even though DDT has been banned from use in the United States for decades, US babies born today are still showing traces of contamination from this chemical. Clearly, we have only scratched the surface with regards to the "long-term" risks of pesticide use and their effects on biota.

While "Frankenseeds" or GMOs (Genetically Modified Organisms) have had a number of well-proven benefits, such as "no-till farming" protecting precious topsoil, they are by no means a free lunch either. But nothing ever is! The use of GMOs, for example, originally decreased the use of herbicides, but also created superweeds, spawned chemical resistant insects, and caused biodiversity declines—specifically the eastern and western monarch butterflies, both currently on the threshold of global extinction. Their dwindling populations have been tied, at least in part to GMOs, but also climate change, habitat loss in the US, and deforestation in Mexico. The list of cross-contamination of chemicals in our food supply is endless and a number of pesticides are either known or probable carcinogens, and all of us are chocked full of them. Once inside a multi-celled organism such as a human, chemicals (both natural and anthropogenic) can trigger cancer in a variety of ways, including disrupting hormones, damaging DNA, inflaming tissues, and turning genes off and on.

While it is true that some cancers have declined, other cancers such as colorectal, thyroid, and testicular cancers along with melanoma and non-Hodgkin lymphoma are on the rise. Some of the increases in these cancers have been attributed to our longer lifespan. But it still begs the question, "Are some of these cancers tied to anthropogenic chemicals as well???"

## Natural Pesticides and Antioxidants

While plants make their own pesticides to protect themselves from insect attacks, many of these natural chemicals can be used as non-anthropogenic repellants. Compounds made from cedar oil are used to discourage moths from eating our clothing. Chemical compounds extracted from the roots and bark of the pawpaw tree have been used as a natural pesticide on garden plants and the chemical compound Acetogenins, from the same tree is currently being used in cancer research.

The poisonous substance known as cyanide is prevalent in both flora and fauna, including such common animals as millipedes. Apple, peach, and cherry seeds contain the compound amygdalin, which when degraded or damaged, release hydrogen cyanide. In fact, chewing several hundred apple seeds is considered a lethal dose in humans. But in small amounts, our liver can process and expel this dangerous poison, leaving no lingering damage. This is a result of a long-term co-evolution with cyanide in our environment.

Furthermore, many of these natural plant compounds are referred to as antioxidants (phytochemicals anthocyanins, flavonoides, and carotenoids), some of which have been shown to suppress harmful free radicals in the human body, and therefore, considered important in the fight against cancer. Several studies have found that organic crops can have anywhere from 18-69 percent more antioxidants than conventional crops. However, when it comes to nutritional differences between organic verse inorganic foods, a wide range of findings have been reported. The fact that many organic farms are plowing the same "depleted soils" that were once chemical farms might explain the discrepancies seen in opposing research. Clearly, there is much exploration left to be done on all chemicals and their effects to humans and the environment, whether it be from natural or anthropogenic sources.

• • • • • • • • • • • • • • • • • • • • • • • • • • • • • • • • • • • • •

**So what's the "rethink" on anthropocentric chemicals?** For me, it's simple. Eliminate or reduce the use of pesticides where you can, especially around the house and yard. Learn to tolerate and live with a certain level of pests, after all, we live with each other "*nature's biggest pest*". Better yet, give them what they want, away from where you don't want them. Rethink the beauty in all things, especially those persistent and terrors of the turf, **DANDELIONS**!

# The Honey Walk Marathon

It all started as a 2-week trek through a nearly impenetrable jungle to a cloud forest located on top of the Maya Mountains. It was like any other day in the jungle where one is usually singing "rain-rain go away, come back another day!" Ever wonder where those other days end up? In southern Belize.....where it rains over 200 inches a year, soaking you to the bone! That's where the term WET BONE comes from, a non-medical condition afflicting biologists that insist on working in the Maya Mountains during the rainy season. Anyway, back to the story at hand. I met up with my Mayan guides who would lead me to the base camp which was about a two-day walk.

While Mayan women are shy and reclusive, Mayan men are bold, proud, and thoroughly enjoy an "ego brawl" —challenging a newcomer whenever, and however, they can. With Mayan men, you must earn their respect, especially if you're an outsider. After a day of trudging through miles of drenched jungle, we came to rest at the base of a hill. As we all caught our breath, I noticed my league of ego-brawlers sucking down a golden brown concoction and asked "What's that you're drinking?" With smirk-filled smiles, they evasively answered, "Raw honey. Do you want some?"

I answered, "No I'm good," which seemed to please them. It wasn't until later that I would come to fully understand those smiles of deception.

Under the veil of their native tongue "Kek'chi Maya" I was unable to understand their scheming conversations to kick my lily-white ass on the hill that lay before us. So unbeknownst to me, a test of staying-power was about to begin. Up climbed the honey-filled Mayans with the enthusiasm of a kid with a new toy (me). Little did they know, as a retired US Forest Service employee, I worked and lived in the Appalachian Mountains and ate hills like this for lunch, that I had walked 1700 miles of the Pacific Crest Trail, or that my nickname was "**Goatslug**".

As they powered up the hill, they kept turning and looking for my punk ass, thinking I would quickly and, shamefully fall behind. Despite their failed efforts to leave me in their dust, (well actually ankle-deep mud), I remained on their heels. I was like a drunken fruit-eating bat on fermented figs.

Walking faster and faster failed to leave the Goatslug of eastern Kentucky behind. Even at twice their age and a belly void of energy-boosting honey, I managed to keep up. We all finally made it to the top where we sat down to catch our breath. No one spoke but me, as I uttered these words. "So guys, *when do we get to the real hills*?" Respect had been earned and the name Goatslug will forever live on in the mountains of the Maya.

By the way, I received the quirky nickname "**Goatslug**" from the fact that when in the woods I walk and climb hills like a goat and while driving a car move along at a slug's pace, hence the name Goatslug!

# Hanging with Fame

When we decided to take that leap of faith and move to Belize to assist BFREE, our mantra was "We were tired of watching the Discovery Channel, we wanted to go experience it." Little did we know then that living in Belize would allow us to befriend two of the best wildlife documentary filmmakers in the business, Richard and Carol Foster. Few things would be more rewarding than our friendship with this extraordinary pair whose films for National Geographic, Discovery Channel, Animal Planet, and BBC received worldwide acclaim including Emmy Awards for dedication and excellence in photography and motion pictures.

The first time we met the dynamic couple was at their house and property in Belize adjacent to the Belize Zoo. Their place looked more like a Hollywood studio—but not the safe soundstages in California for overpaid actors—but film sets for underappreciated animals in the wild jungles of Belize. One of the most famous sets was the one used to film a "Jesus Christ Lizard" known for its ability to run across water as part of a National Geographic Project, a 5-second bit of footage most of us have seen while sitting at home in our recliner chairs.

Visits were often spent listening to commentaries about their filming as we watched their adventures unfold on screen. Stories like the time Richard was filming a whale shark that nearly swallowed him alive while the shark attempted to suck in a mouthful of krill (it's preferred food, not Richard) or the time Richard had to keep pulling a capybara carcass back out of the jungle to film a mountain lion in Venezuela.

Perhaps the most unusual conversation I have ever had occurred while at the Fosters eating dinner one evening. Richard, a guy with an insatiable sweet tooth, abruptly stood as soon as the meal was finished and casually announced, "As much as I would love to stay and have dessert with you, I need to go feed the vampire bats." That statement by itself would startle the ordinary guest but we took it in stride. After all, we had witnessed firsthand the Foster's dedication to the craft of wildlife documentary filming. But the next part would even catch us off guard!

Being the curious biologist that I am, I asked if I could come along. Normally, Richard was eager to share his work with me but this evening he said no. Why? Because the normally skittish bats would not be fed the blood of swine as he had previously done when working with these intelligent creatures—but the blood of Richard!

In preparation for a film sequence for BBC about Manuel Noriega and the dreadful prison system in Panama during the 1980s, he first trained the bats to feed on pigs and was now training them to feed on a human and he was the human! To ensure accuracy, Richard constructed a mock prison cell to mimic the poor conditions of the prison where vampire bats fed on the prisoners' blood in the middle of the night. Richard would return a few hours later, bleeding profusely from the bat bites whose saliva acts as an anti-coagulant but grinning from ear to ear. "I finally got the bats to feed on me!" he announced proudly. While that would send shivers up most people's spines, I thought it was fascinating. "Obviously!" and we all laughed at the sight of a blood-soaked Richard. Talk about unconditional commitment to a film!

But Richard was not the only badass of this team! While filming "Land of the Anaconda" in Venezuela, Carol would wake up in the middle of the night with a sharp pain down her spine, unable to walk! She would be airlifted out of the Llanos to the nearest hospital where she received a diagnosis of Multiple Sclerosis that would have been debilitating and disabling for most ordinary mortals. But not Carol, the queen of badassery! After only ten days, still unable to walk, she returned to the remote filming location in the Llanos, where her tiny frame was carried to the swamps for filming on the back of a bulky Venezuelan. There she would film the infamous scientist, Jesus Rivera, who used his bare feet to locate the gargantuan snakes. What an inspiration!

We would spend many years with the Fosters, assisting with filming around BFREE on various projects including "**The Rainforest is Our Classroom**" which can be viewed on YouTube along with numerous other films they made. I was even able to spark their interest in land snails, not an easy task! Richard constructed sets to capture never-before-filmed interactions between *Sibon* snakes (snailsuckers) and native Belizean land snails. In acknowledgement of their significant contributions to the world of conservation through film, I named a land snail new to science, in their honor. Rosy Marauder, *Euglandina fosteri* (pictured here) is endemic to Belize, found nowhere else in the world.

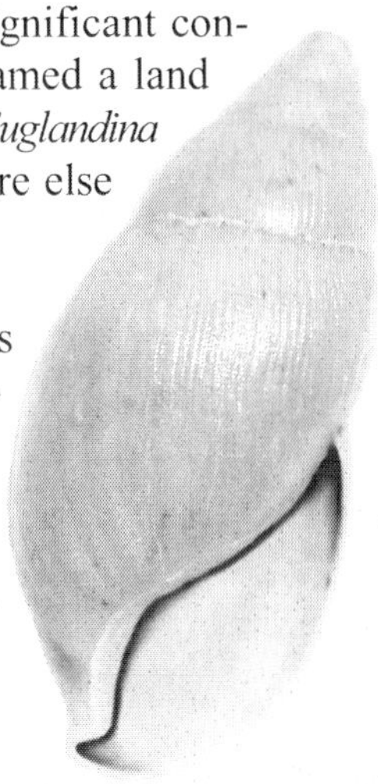

Ironically, for all the precarious situations that the Fosters encountered while filming some of the world's most dangerous wildlife, Richard would meet an untimely death at the hands of the most dangerous animals of all, *Homo sapiens.* We remain dear friends of his widow and colleague, Carol, and cherish the memories of this extraordinary couple, *living where the cats eat the dogs*.

# “Num se chukka ick”

One of my better memories while at BFREE would be working with our Mayan staff, especially the guys. Not only did they speak their native tongue of Ket'chi, but English as well. Nevertheless, there remained some interesting social language barriers—some that we had fun with and some that were downright embarrassing for me. One example was the word "**punk**". When I first came to Belize, I teased the younger guys by calling them all punks (just like I did to any young person back in the USA). My teasing was always met with an uncomfortable silence until one day Marcelino Pop pulled me aside and asked me if I knew what that word meant.

*I said, "Sure a delinquent kid of sorts."*

"It may mean an alcoholic kid where you're from, but in Belize, it means you're gay and to many Belizean men, this is a monumental insult," Marcelino replied.

*"Oops, sorry, my bad!"*

In a nod to cultural differences, I started calling the guys "**bunks**" instead which they were all cool with.

While Maya women don’t care much for joking and just turn away, Mayan guys take pleasure in it, so I had to constantly keep on my toes. Like the time Tom Pop put a live 6-inch scorpion on the truck seat, minus the stinger he had first removed. Imagine my reaction as it crawled over my neck!

Marcelino and I would often exchange one-liners such as “**If you don’t stop that, you’ll go blind**”, a line I often used when I caught him sneaking sinister looks at the lady cooks. At first, not knowing me all that well, I think he partly believed it but as all good jokes must play themselves out, he decided I was mostly just full of shit.

Another one of my all-time favorite one-liners was from my dad that I think he got while in the Navy. He use to say it to us kids when we were naughty; “**Seema-Laka-Chu”** which according to my dad meant, “*Death shall be your reward*”. Marcelino would come to love this one and use it often on his own kids at home.

When in Kentucky, I picked up many different one-liners such as this one: “**I’ll smack you upside top across your pealong”** the *pealong* meaning your head. It was one I used often and one that Marcelino would also add to his Mayan vocabulary.

Last, but certainly not least, I was determined to learn at least one line in Kek'chi Maya before returning home to the states which ended up being "**Num se chukka ick**" meaning "pass me the hot sauce".

Now my idea of hot sauce, or I should say my tongue's idea of hot sauce, is ketchup. I had spent most of my seven years in Belize avoiding the tongue-burning Marie Sharp's that Belizeans slathered on every bit of food that passed their lips. The fact that this was the only phrase I would learn in seven years of living with the Maya was ironic, to say the least!

# Subcutaneous Horrors

What's worse than botflies living under your skin? Flesh-eating parasites dining on your paw! While on a return visit back in the states, I noticed I had a small volcano-like wound on my left hand that refused to heal. Thinking it was a botfly, I covered it with duct tape to kill the unwanted larva. This typically takes around 24 hours or less but after removing the tape and waiting a few days, the ever-enlarging crater was not going away.

This didn't make much sense, since I had dealt with dozens of botfly assaults on my body before. It was at this point I started to think, *Maybe it's not a botfly but something more hideous in nature*. As it turns out, it was a flesh-eating parasite referred to as "***Cutaneous leishmaniasis***" caused by a single-celled protozoan transmitted by the bite of a phlebotomine sandfly. It was slowly eating my subcutaneous flesh away, not a very pleasant thought or sight.

There wasn't much pain except for a small burning sensation at the site. I had seen what *leishmaniasis* does to people in Belize who have lost parts of their ears and face to this wicked parasite. Since I am rather fond of my bodily parts, the thought of being reshaped by this microscopic organism was daunting, to say the least. The standard treatment for American *Cutaneous leishmaniasis* is somewhat controversial, using Pentavalent antimonial drugs (sodium stibogluconante and meglumine antimonite) heavy metals that have been used since the 1940s. However, these treatments are expensive, toxic, and painful to use.

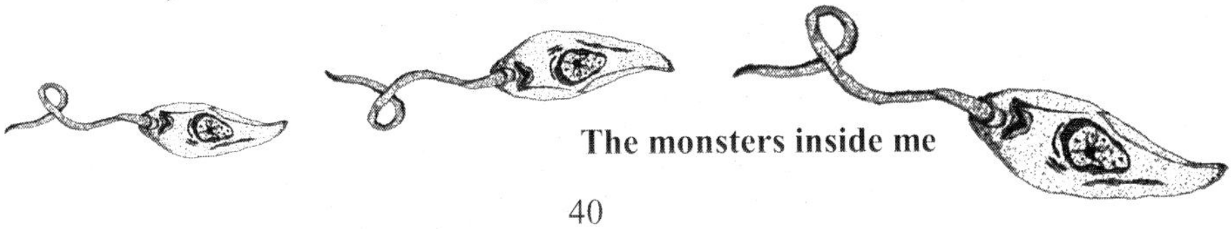

**The monsters inside me**

I decided to go with a less poisonous method, which although controversial as well, had shown positive results. The treatment went as follows: I applied wet towels as hot as I could stand to the site 3 times a day for around a week. Fortunately, the treatment was 100% effective at killing the parasite. According to what I read, the advantage of hot treatments over toxic metal therapy is that a natural immunity develops in your body (I never got it again) and you avoid the heavy metals which are reported to have long-term, unhealthy side effects. The only thing that remains is a small scar and a slight phantom pain at the site which I don't fully understand. It, however, remains a constant reminder of this fascinating but freakish parasite.

# The Cat Whisperer

My first big cat encounter was at BFREE where I ran smack into a mountain lion one late night while on my bicycle. It left us both terrified and bewildered! The cat ran away and I got back on my bike and continued on. I went from cat rear-ender to cat whisperer overnight and not the "pussy cat whispers" of reality TV. I'm more of a "cat teaser" and the bait I used to lure cats in from the surrounding jungle was a high-pitched distress call made with my closed fist. To a curious and hungry cat, it's an irresistible noise that I have used successfully to call in all five native cat species in Belize. It also works on brocket deer, peccaries, and foxes.

Rebecca, an intern at BFREE for nearly six months, was desperate to see a jaguar before returning home to the states. Night after night, we searched in vain, hoping for just a glimpse of one of these magnificent cats, but mostly seeing only nightwalkers (local name for kinkajou) and gibnuts (local name for paca). On one of her last nights at BFREE, I said "Let's go bag you a jaguar memory." So off we wandered into the dark and foggy night to find the elusive cat. In the jungle, there are dark nights where you can't see shit without a flashlight and then there are those ludicrously dark moonless nights you can't see shit if you fell face down in it. This extraordinary night was the latter.

As I was about to step in a big pile of tapir dung—thank technology for headlamps—we heard a male jaguar woofing (calling) for females—not an uncommon sound during mating season. There in the distance, we could see the unmistakable yellow eye-shine of the king of the jungle.

The "cat teaser" sprung into action as I started to blow through my closed fist creating those squeals of distress, promptly getting the attention of the curious feline not more than 100 yards away. As we both hunkered down, I continued to make those closed-fist calls. It worked! The inquisitive jaguar moved closer and closer until we could hear the cat's heart thumping and heavy breathing. *Wait! That was our hearts pounding not the cat's!* As you might imagine, we were having a close encounter of the paramount kind as our emotions went through the roof—well, actually the jungle canopy. And then reality hit me like a "ton of dumb". *What was I thinking?* I had just called in a 175-pound predator that's both horny and hungry and was now, just leaps away.

Clearly, I had let a feeling of awe cloud my better judgment when it came to this possibly dangerous animal encounter. But as luck would have it, the curious but cautious jaguar was smarter than I was that night—seeing my phony distress calls as nothing more than an annoying diversion from its nightly prowl. The big cat—now bored with my antics—simply turned and disappeared into the mist. But Rebecca had finally bagged her jaguar experience in Belize. One neither of us will ever forget.

# The Meaning of Respect

Just another soggy day expedition in the jungles of Belize or was it? From the depths of the rain forest and miles from BFREE came a challenge of trunk and limb that would take my relationship with Tom Pop to a whole new level of respect and friendship.

After an arduous day of crawling down land snails and the sore knees to prove it, base camp was usually filled with laughter and raucous conversation with questions like "why would anybody study snails?"

My response was quite simple really. "I figure when I'm 90 years old, I'll still be able to catch any snail on the planet." A response that usually brought fits of laughter from our Belizean staff.

But this evening would be a test of strength and character. Tom, who is half my age and built like a Ceiba Tree, challenged me—a skinny, white, middle-aged guy from Kentucky—to a test of just plain old-fashioned "pushups". Tom had seen me "pumping flesh" around BFREE to stay in shape and thought he was ready to take me on and teach my scrawny-ass "the meaning of respect. "

"Tom, are you sure you want to do this? I generally keep in shape by doing around 100 pushups every other day." Tom didn't care and was ready and anxious to kick my butt, so under the watchful eyes of spider monkeys and a few other researchers, the ground-hugging battle of the arms began. The onlookers counted.

Five...Ten...Fifteen...Twenty...Twenty-five...Thirty...Thirty-one...Thirty-two...

And then, a two-second pause for Tom, but I kept pushing my muscle fibers to the limits. Tom was beginning to tire but would have no part of an ass-licking in the jungle by an ass-kicking old fart from Kentucky.

Tom struggled on with a few more arm-flaming pushups until suddenly his arms just couldn't push anymore. His biceps now trembling from fatigue, he finally collapsed, and it was humble pie for dinner.

I continued.

Forty...Fifty...Sixty...just enough to teach Tom the true *meaning of respect...*all caught on tape by another researcher.

I remember saying during the arm-burning ordeal *You're not going to let an old man beat you are you?*

The ironic thing is that I've seen Tom drag logs through the jungle that would destroy me. It's really just a matter of repetitive training which gave me the edge on Tom when it came to pushups. In any case, I had earned Tom's respect and he had earned mine, for being audacious enough to challenge someone he barely knew and, in front of others he knew not at all.

## The Lost Temple of Camazotz

High pitch sounds emanating from deep within an endless jungle would lure an elite group of explorers searching for "**Camazotz**", the lost Bat God of the Maya. The voyagers included Mark Gumbert (the leader), Price Sewell, Tom Pop, and myself. It would be a two-day journey by foot through snake-infested jungles and relentless heat to Quebrada de Oro, the Maya name for "Valley of Gold". Hence, we hiked forward with "dreams of riches"—wait we are bat biologists, not gold-diggers.

*What I meant to say was "dreams of never-ending bat encounters.*

Bats are a fascinating group of animals that comprise approximately one-quarter of all mammal species on Earth. Elusive, highly evolved, and bizarre are the hallmarks of these winged fur balls. We would spend the next two weeks deciphering the secrecy of bats attempting to answer the questions: What kinds of bats live in the Maya Mountains? What do they eat? Where do they live?

The process of surveying for bats is fairly simple. String up fine mist nets between two tall poles in locations where bats frequent such as over streams or in forest corridors. Then wait for the unsuspecting flying mammals to become ensnared in the nets and hope you get to the bat before a hungry margay does.

While most bats can sense and avoid these nets, there are those bats that just aren't paying close attention and soon become entangled in "Gumbert's Web'. It's similar to that clueless motorist that ends up in a ditch because they were "texting" somebody that Joe's Bar and Grill forgot to stick one of those fancy toothpicks in their Double-Decker hamburger.

Once a bat is ensnared, a tiny transmitter is glued to its back with mammal-safe glue and the bat is then released. The next day, we all headed out following the signal from the bat by use of a telemetry device.

On this day, the signal took us to the base of a 500-year-old hollow fig tree. Scanning the site, the crew began recording data like tree size and surrounding habitat.

But wait—what was that cone-shaped hill doing here in the middle of a otherwise level floodplain? A closer look revealed an incredible sight. We had just stumbled upon an ancient Maya Ruin! *Was this the city of gold? No such luck!!* But it was nothing less than astonishing.

The bat had guided us to a lost city in the middle of the Maya Mountains. There were tombs, ball courts, and multiple other structures all covered up in jungle. It was like a clip from an Indiana Jones movie, but the real deal. We spent the day in this lost metropolitan city of the Maya imagining what it might have been like a thousand years ago when the Mayan bat God **Camazotz** ruled the night skies in the city of gold.

# Chinch's, Botflies, and Scorpions, O My

## Chinch of Terror

The most dangerous animal in the jungle is not the jaguar nor the fer-de-lance! That distinction belongs to an inch-long insect known as the *kissing bug* (pictured above) or locally referred to as "chinch". A member of the Reduviidae family, kissing bugs feed on the blood of small mammals such as mice and bats, but if given the opportunity will also feed on larger mammals, including humans. These "true bugs" often feed around the mouth where the blood supply is abundant, hence the name *kissing bug.* Of major medical significance in the New World tropics, several species in this family transmit Chagas disease, a debilitating and often fatal syndrome caused by the protozoan *Trypanosoma* (*Schizotrypanum*) *cruzi.* Chagas disease has infected 7 million people and kills around 10,000 a year, far more than snakes or jaguars.

The protozoan is not introduced through the bug's proboscis (biting apparatus) but instead, the parasite resides in fecal material left on the skin surface during and after blood engorgement. The protozoan enters the host's body (generally through the bite, but sometimes around the eyes) when the host scratches the itch caused by the bug's bite. Once inside, the parasites multiply rapidly and congregate in vital organs, including the heart. If promptly attended to within the first three to four months of inoculation, the disease can be successfully treated with antibiotics, leaving no lingering effects. It is speculated that Charles Darwin (a famous naturalist) died of Chagas disease.

So for us, bites from the kissing bug, not jaguars, were always of major concern, living in the jungle *where the cats eat the dogs*. These nocturnal bugs were everywhere, in caves, in your house, even in your bed. We, however, always slept in the safety of a bed tent which would not only keep out kissing bugs but also scorpions, ticks, and spiders.

## Flies of Repulsion

If the most dangerous animal in the jungle is the kissing bug, the most disgusting animal is probably the botfly or more precisely, its larvae.

An adult female botfly with eggs is indeed a devious creature with no direct deposit on the host (any number of animal species including humans). Instead, the fly commandeers other insects to carry its eggs to the host, choosing to lay its eggs, instead, on bloodthirsty insects like mosquitoes or doctor flies (deerflies). A mosquito with hitchhiking botfly eggs on board then seeks out a blood meal. Once the mosquito lands on the host animal to feed, the botfly larva hatches and crawls down the leg of the mosquito and immediately burrows into the host's skin.

The botfly larva goes through several instar (life) stages growing in size, feeding subcutaneously, and forming a breathing port to the outside to satisfy its increasing oxygen needs. A fully mature botfly larva with spiracles through which it breathes in its last instar crawls out of the host (similar to the scene in Alien), drops to the ground, and pupates; thus starting its amazing life cycle all over again. In the early stages of parasitism, the larva is easily killed by placing duct tape over the breathing port for 24 hours. Once deceased, the larva can be squeezed out or if small enough, left for the body to reabsorb. Fortunately, the parasite does not carry any known infectious diseases.

While the botfly larva is disgusting to most humans, there are a few of us biologists that think botflies have one of the most fascinating life cycles of any animal. In fact, I know several biologists (I won't mention names) that allowed the larva to go full term to experience giving birth to an alien invader. One such person had a botfly larva embedded in his forehead that went to full term, "giving birth" while conducting a bank transaction—you know—at the bank. It dropped from his swollen head right in front of the teller's window along with no shortage of disgusting bodily fluids. As you might imagine, the freak show was not soon forgotten and, to this day, likely haunts that teller's dreams.

During the 7 years at BFREE, I provided a warm and secure home in my subcutaneous tissue for hundreds of botflies. I quickly came to recognize the larva's presence however by the needlelike digs inflected while chewing on my flesh. It's not as gross as you might think though, and I learned to promptly terminate the unwanted invaders during the first trimester by the use of Duct tape. As time went on, my body somehow learned to kill them naturally. If you live in a healthy jungle environment, botflies are just a way of life and people here don't give them a second thought.

**Knee Buckling Pain of Ek Chuaj**

The subject of discussion here falls on one knee-buckling species of arachnid, the 6-inch long forest scorpion, or as it is known to the Maya,

**Ek Chuaj**. When we first arrived at BFREE, we were assigned an unoccupied cabin—uninhabited by humans that is. The staff had done a great job cleaning and, as far as we could tell, the space was free of creepy-crawlies—little did we know! It wouldn't be long until we saw our first 8-legged pain-givers of darkness.

So each night, I would catch and release scorpions to the outside unharmed—after all I didn't want to kill them! Dim-witted me. By the end of the first 6 months, I had removed more than 75 scorpions from the house. Where were they all coming from? Perhaps they were the same ones I had spared returning home?

My tolerance for scorpions in the house was about to change forever. One painful day while cleaning out my backpack, I felt a burn on my fingertip. Surprised at first, I wondered what the hell that was. As I shook out my pack, a 6-inch long scorpion fell to the ground and scurried away, leaving only the pain of its sting.

*Ouch, that hurts, no that really hurts, mother fucker, son of a bitch my hand felt like it was going to explode!* So what's it feel like you ask? It's excruciating, knee-buckling pain, like being stung by an 80- pound hornet or someone taking a sledgehammer and mashing your finger flat! I was close to crying like a newborn baby. On a pain scale from 1 through 10, a solid 11.

You've got to respect something that can deliver such pain from one tiny drop of venom. Within 30 minutes my finger and hand went completely numb and by the next morning my tongue as well, making it weird to talk. But no other side effects resulted with the exception of newfound respect for the forest scorpions of Belize.

For Judy, her scorpion experience would be much later. In fact, for the next few years, she would annoyingly brag that she had escaped the wrath of the nocturnal stinger. But few people that live in Belize long enough do. She would get hers soon enough!

It happened one dark night in our final year at BFREE. Carelessly stepping out of her bed tent without the obligatory flashlight, her foot landed directly on top of one! I heard a scream and knew exactly what had happened. While there was a small part of me that felt empathy, my Asperger's side was saying, "I told you not to test fate!"

She had now joined the "throng of the stung", an elite group of pain-seekers in Belize.

# Cum All Ye Men to Church

*There are churches in Belize to soothe the savage desires of men and churches in Belize to rescue their lust-filled souls.*

Personally, I think a better use for churches would be to turn them all into homeless shelters. I'll admit upfront I haven't been to a church for a really long time except for those occasional, outrageously-expensive, coupling events of two overly zealous individuals.

Nevertheless, one day, I overheard Tom and his brother Marcelino, our two Mayan staff members, talking about their recent religious experiences at church; with one of them proclaiming if he could, he would go to church every day of the week.

*"Wow! You guys really take your church services seriously don't you?" I remarked.*

"Yes!" they both replied without hesitation. "The services we get at our church are heavenly."

*"Heavenly?" I asked.*

"Yep, the people we meet at church are always friendly and make you feel right at home. It's like a family away from family. You never leave church disappointed or if you do it's because you *came* at the wrong time," one eagerly responded.

Now, I have been to a number of different churches but never a Mayan one, so curiosity and naughty thoughts were beginning to flood my mind.

*"Is it a Christian-based religion?" I asked, part serious, part kidding.*

"It's a religion that provides a spiritual union for all who truly believe and come in the house of the Lord!"

*"So what are the basic doctrines?"*

"Well "love" of course, silly boy!"

*"That's a great philosophy for any religion," I responded.*

Tom continued, "If you're a man feeling lonely and unloved or just plain bored with your home life, then this is the church for you."

*"How about women and children?"*

"Children are strictly forbidden but there is no shortage of fine-looking ladies, my favorite being Lacy Underall. She's always at the head of all who come to the congregation," Tom stated emphatically. The story was reaching a climax.

"Hey, we'll be going tonight after work. Why don't you join us?"

Finally, the cat was out of the bag and their innuendo was beginning to register—my immoral curiosity had just struck lust!!

This was not a church of God but a dwelling for manly expressions of biochemical desires. Not a church of love but a church of lust, a church of ill-repute for men bored with their same old faithful wives.

*"I'm agnostic so I'll pass," trying to deflect their request since I was not interested in visiting a brothel.*

Tom responded, "I don't care how you eat your eggs, Dan. I guarantee you will have a good time if you come in our church, but you can't tell Miss Judy, she wouldn't understand."

*"You're right about that Tom! She's funny about deceitful and two-timing husbands since the two before me cheated."*

Tom's persistence would continue for months. I guess I should have considered it an honor to be invited to run with the Mayan pack. In retrospect, perhaps out of respect to the Maya Culture, I should have taken one for the team— just kidding, Judy—mostly???

Tom refused to take no for an answer.

*"I'll tell you what Tom, if Miss Judy is ok with me committing adultery, I'll make an effort to cum at your "Church of the Unfaithful"! But you need to ask her"* —knowing full well beforehand what her answer would be.

It wasn't the answer Tom and I clandestinely hoped for—every man's dream right? A free pass? This, however, would end Tom's endless nagging to get me to commit betrayal—so curiosity and my wife didn't kill this faithful cat!

# Snack Attack

One fine day, Ron Caldwell, a good friend and colleague from Lincoln Memorial University in Tennessee came to BFREE to help me conduct research on that most exhilarating group of organisms, land snails. Since BFREE was already full of visiting professors and students, Ron had to stay with Judy and me at our tiny jungle home. Ron, a generously proportioned man who enjoys eating large quantities of delicious food, would quickly and quietly become bored with the fine cuisine of BFREE.

But while BFREE does provide enough food to fill the belly, victuals here can get a bit boring at times, ok really boring in fact, absolutely in the depths of despair boring! Let's face it there are only so many ways you can fix rice and beans or beans and rice or rice with beans—you get the picture.

Ron's answer to the food diversity shortage was to sneak in cases of irradiated bacon with a half shelf-life of 500 years, chemically stable hotdogs, and endless varieties of snacks (potato chips, candy bars, and beef sticks).

As you might imagine, the Maya staff, especially the guys, quickly befriended him not for his irrespirable charm but his stockpile of endless tongue-pleasing treats. I must admit I also enjoyed Ron's basket of pleasures at times but don't tell anyone.

One late night, the sound of gnawing woke me from a nightmare of killer beans attacking a bowl of benevolent rice! As I lay there happy to be free of that ghastly dream, the sounds of crunching and gnawing continued into the night. I turned to Judy, now awake from the noise.

"What is that's chewing noise?" she asked.

*"I think we have a vespers rat in the house, one of a dozen or so native rat species here in the jungle."*

The next morning would reveal the culprit. There on the floor lay piles of candy wrappers, empty cans of Vienna sausages, and a few cracker crumbs. Our visitor wasn't a vespers rat at all but a *Homo sapiens* on a late-night feeding binge of junk food proportions. This would happen night after night until the "stacks of snacks" were completely and thoroughly consumed. At that point, there would be nothing else to do but restock the stash of unhealthy, corpse-preserving munchies, and the legend of Snack-attack Ron would live on in the bellies of men and rats alike.

## You Can Run But You Can't Hide

Suddenly, there's an eerie silence among the trees followed by light rain falling on the jungle floor—but no rain. It sounds like the pitter-patter of tiny feet—but nothing in sight. Appearing out of nowhere come swarms of flying insects, crawling snakes, and climbing salamanders. What's going on out there? Then I see it—moving along the jungle floor like a tsunami of biblical proportions yet only a quarter-inch high. A song begins to play in my head ♪ *the ants go marching one by one hoorah, hoorah* ♪. Not the might of one but the might of many, army ants by the tens of thousands were advancing across the jungle floor, driving every living thing from its home. The evacuees were buzzing, chirping, and screaming "flee for your lives" but few would escape the marauding butcher machine. Army ants are indeed a force to reckon with—so voracious, I have seen them take down Africanized honeybee colonies.

Army ants build no permanent nest, but instead form living bivouacs made of interlocking live ants where the eggs, larvae, and queen are cared for and protected. Each day, a raiding party leaves the bivouac to pursue fresh meat which includes mostly invertebrates. Prey is caught, killed, and butchered onsite then carried back to the colony. Several bird species including antthrush, brown jay, and woodcreeper follow these ant raids from a safe distance, gleaning quarry that escapes the marauding ants.

If you live in the rain forest, however, they will eventually come knocking at your door. You can step on them, spray them, or nuke them but they just keep coming. The ants are unrelenting. Surprisingly, instead of trying to beat them back, people in Belize welcome army ants into their homes with open arms. Locals just leave their dwellings until the ants are gone, typically within a few hours of arrival. Our jungle home fell in the path of several army ant invasions and it was nothing less than mind-blowing to witness.

Within minutes, our home was covered with thousands of ants. They were everywhere—in cabinets, dishes, on the walls, and ceiling but thankfully, not in our zipped-up bed tents. Before our very eyes, ants were catching, killing, and dismantling kissing bugs (*hooray*), scorpions (*don't miss these stinging devils in the house),* spiders (*do miss these*

*critters*), and any other multi-celled animals that got in their way. Carrying the fresh kills in their mandibles, the ants eventually headed back to their hive. Within an hour or so they were all gone—not an ant or bug to be seen anywhere in the house. It's a little known fact that ants are extremely important custodians of balance in forest eco-systems; a single tree containing as many as 40 species. Nature's natural insecticide but without the nasty and dangerous chemicals. How cool is that?

# Poetic Justice Served Jungle Style

Ever wonder what happens to a domestic cat that messes with "poetic justice" in the jungle? This is a true story about what went terribly wrong for one such feline. The names have been changed to protect the innocent—wait just a minute—there's no virtue here, only jungle justice.

Squishy the cat, had the good life at BFREE—one square a day but most importantly, a safe and sound place to go to at night, or so we thought. As it turns out, it doesn't matter if you're inside or outside, the Belizean jungle is no place for a pussy cat.

When the Marlin family left for the states, we were put in charge of Sofia's pet cat, Squishy. Feeling sorry for the caged-up feline, we decided to let her wander the research facility during the day, and at night, lock her up for safekeeping, or so we thought. Squishy could come and go as she pleased by way of an outdoor cage and door attached to the house that led to a bedroom with a closed door, a kind of cat door with a lock. At night, nothing could get in and Squishy couldn't get out to wander the jungle. Remember when I said that the jungle sleeps during the day? Well...NOT EXACTLY! There's always an exception to the rule and this was one of those times the rules were broken.

One morning while on our way to an early breakfast, we noticed Squishy laying face down in the outer cage part of her enclosure. Upon closer examination, she was not only dead but had mucus covering a good portion of her upper body and head. The door was closed and locked. Nothing could get in. As you might imagine we were befuddled. I turned to Judy, "Let's go eat and figure this out later," I remarked. This is when my limited CSI skills began to kick in. I love a good animal mystery.

The night before, Judy and Steven had treated some disgusting botfly lesions on Squishy's backside with some dreadfully strong insecticide. Perhaps she had licked her wounds and ingested some of the chemical, causing a fatal reaction? That might explain the mucus mystery. But her whole upper body was distorted, challenging the poisoning theory. I continued speculating on other possibilities.

What if a snake crawled into the enclosure during the daytime when the door is usually open? If so, then the snake would still be inside the bedroom, since the inside bedroom door is also still closed. After a few minutes of searching, the strangling serpent was discovered—a 4-foot long boa constrictor hiding behind the cabinet!

Apparently, the snake had crawled through the open door sometime during the day, unbeknownst to us and the cat. That night, we inadvertently locked Squishy up with the snake, during which time the snake caught and squeezed the life out of the feline. Squishy was on her way down the snake's mouth but the snake was too small to completely swallow the larger prey and therefore, was forced to regurgitate its intended meal. This explained both the mucus surrounding the cat's head and the distorted body. Squishy had just spent all 9 lives in one unfortunate night. *The blood was on our hands!*

You are probably wondering how this was poetic justice in the jungle? The night before her untimely death, Squishy had caught and killed a small coffee snake just inches from where Squishy was squashed by the boa constrictor.

• • • • • • • • • • • • • • • • • • • • • • • • • • • • • • • • • • • • •

**A Global Quandary!**

Ok, jump on the educational bandwagon with me for a moment. Here are some sobering facts about domestic cat damage on wildlife populations. In the US alone, domestic cats kill between one and four billion birds a year, as well as 6.3 to 22.3 billion small mammals and hundreds of millions of reptiles, amphibians, and arthropods. Across the globe, the numbers are unknown. While many cat owners insist this is just the cat's natural behavior, it's really not. In general, wild felines chase, kill, and eat their prey but when their bellies are full, stop hunting. Domestic cats, however, are caught between two worlds—one where they are well-fed and one where they are still driven by instinct to hunt. With full bellies, domestic cats will still hunt and kill but often don't eat their quarry. After the prey becomes lifeless, the bored cat goes on hunting and killing more and more wildlife, leaving a wake of broken and dead animals behind.

# Curiosity Doesn’t Always Kill the Cat

The wails of jaguars and flashes of iridescent blue butterflies are the very essence of a Belizean rain forest. For some, a mere glimpse of the big cat stirs excitement within—for others a way to feed their family. Curiosity doesn't always kill the cat—but apathy often does! For on that unfortunate day, earth was pushed ever closer to "**biodiversity-genocide**". Off in the distance we could see the all to common circle of vultures that feed on death. We soon found the beheaded carcass laying still on the jungle floor—where just hours before, the cat stood proud. The once majestic feline had been slain for its teeth and sold to naive tourist for mere pennies. The carcass discarded like rubbish.

We are currently on course to end the continuation of half of all life on earth—which by 2100—could be as many as 5 million plant and animal species. Our greatest moral dilemma and inadequacy will always be the collective failure to consider the intrinsic values of all things great and small. Until we stop monetizing our natural resources and treating them like a commodity, we will burn through species like a hippy burns through cannabis. Instead, we need to place them with our greatest labors of time—the Mona Lisa—for they too, are priceless and irreplaceable gifts. As Aldo Leopold so elegantly wrote, *“We stand guard over works of art, but species representing the work of eons are stolen from under our noses”.*

In Belize, as elsewhere, many animals are killed for body parts, sold to tourists, or shipped to Chinese markets where canines are worn as symbols of power or used in unsubstantiated Chinese medicine. Like overcompensating Trump Brothers who hang their trophy egos on elephant ivory or wipe their dirty feet on leopard skins, the purchase of jaguar teeth is the trigger that kills the cat. What sad endings to such magnificent toils of evolution! By simply refusing to buy illicit animal parts in places like Belize, you disrupt the circle of transgressions. If there's no demand, there’s no slaughter.

As a biologist who is also an atheist, I am sometimes asked, "If there's no God and the universe doesn't care if we live or die, why should we care if jaguars are killed or butterflies disappear tomorrow?” It’s a great question! If I was a chimpanzee with limited thinking, it would probably matter little, if at all. But chained to our human brains are complex thinking machines that take us past the notion that "it's great to be alive", too "why are we alive"?

Philosophically put "purpose is a consequence of life". Scientifically put, the universe, by way of exploding stars, produced a planet where random piles of incoherent and formless chemicals came together to form cohesive "**life**" (abiogenesis) which through perpetuity evolved into what we are today. Moreover, and in accordance with the laws of thermodynamics, energy cannot be created nor destroyed. In death, the collection of atoms of which everything is composed is simply repurposed—whether it be in the form of a one-cell protozoa or a great blue whale—life emerges from death and continues onward. So we are literally a part of everything around us, as is everything a part of us (reincarnation in its simplest form). So its begs the question, do *Homo sapiens* give consciousness to the universe—just as the sun gives life to the earth? Perhaps in a quixotic way!

Nevertheless, for this awe-inspiring act of astronomical proportions, I consider myself among the privileged and therefore grateful to be alive. Being anything else would be ingratitude and nonsensical! *This alone makes my life worth living and gives me purpose.* Further acknowledging that my life is a result of an infinite number of chance encounters, from colliding meteorites to the consequence of mate selection. Or as Richard Dawkins put it, "*We are born by chance, and our lives are shaped by chance events*".

Even as most would presuppose "purpose" as a result of a **God**, I would presuppose "purpose" as a result of **life**. In other words, from the moment we take our first breath, there's intention through which purpose is established. This could be said for all living things. For most animals "purpose" is primarily driven by instinctive behavior and the limitations or demands placed on them from the surrounding environment. But for humans "purpose" is tied to both instinctive behavior and a higher consciousness of self, and greater thinking that follows. This phenomenon coupled with technological advances has allowed us to easily operate outside the constraints of just "raw animal instinct". As a result, we get to "think and choose" a more philosophical path in our lives than just eating, sleeping, and screwing. It is that "choice", absent the chains of inherent behavior, that takes *Homo sapiens* to that next level of existence, where a deeper meaning of purpose can be contemplated and expressed.

So in my dyslexic-autistic-brain, life—absent a God—facilitates a journey of our own making, as opposed to a prescriptive one from a 2500-year-old book or an impartial universe. Furthermore, protecting all things, great and small is the logical thing to do. Why? Because everything that follows in our footsteps is then provided a place to experience that same extraordinary privilege of life, however brief it might be. This construct alone gives all "life" purpose—even though the universe is without an ultimate or predestined goal.

***In other words, in a "universe" that is entirely without purpose, we can choose a life that is entirely purposeful.***

So for me, a world without jaguars and butterflies is indeed a lonelier place to live. Earth and its perpetual beauty, from ocean bottoms to the mountaintops, from snails to whales, simply enriches our lives in ways we can't even imagine. For John Muir once wrote, "*There is a love of wild nature in everybody, an ancient mother-love showing itself whether recognized or not and however covered by cares and duties.*"

# Naked But Not Afraid

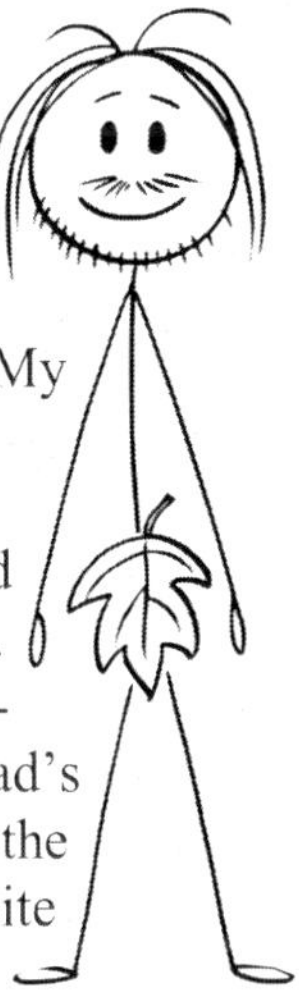

There's normal and then there is yours truly. As stated earlier, my mother noted my strange mannerisms in the early weeks after my delivery. Little did she know just how true her observations would play out and where they would eventually lead me in life. My dad, being a fishermen would say, "*Can we throw this one back*"?

But they kept me anyway and put up with my many strange and offbeat antics for a lifetime, for normalcy was never in my blood. Fortunately, as time went on, my parents would spawn four additional planet-nibblers, (Malia, David, Juli, and John) and to my dad's delight, all quite normal when compared to me. I would become the black sheep of the family, which was a far better fit than the white sheep of ordinary.

I am often referred to as Dr. Doolittle by my friends, but unlike Doolittle who preferred the furry animals, I have always favored those little hairless creatures with exoskeletons. After all, bugs were the first creatures I caught as a kid. Whereas most people get a warm fuzzy feeling from cats and dogs, mine was chasing creepy crawlies. This preoccupation with arthropods would take me to a whole new level of bliss in the jungles of Belize.

It was nearing the end of the dry season in late May when the relentless rains begin to fall—and fall they did. This is the time of year when the jungle is reborn with the passion of a Cat 5 hurricane!

Suddenly, without warning, an explosion of monumental proportions filled the night air with billions of emerging termite queens or flood-flies—as they're called in Belize. As the winged creatures took flight, the air became a cloud of fluttering insects seeking new beginnings. So dense was their numbers you could barely see more than 10 feet away.

They were everywhere including your home, your clothes, your hair. There was no escape! So what does a Aspie-Rogue-Biologist from Kentucky do? Only the most illogical thing I could think of at the time—take off all my clothes to experience the force of countless bugs crawling over my "Naked but not Afraid" body.

As they say, if you can't beat them, the only reasonable thing left to do is to merge with the herd. So that's what I did.

The feeling of tens of thousands of tiny legs and beating wings across my unclothed skin was a corporal sensation beyond words!

While my wife looked on in horror, I reveled in the ultimate sensation. Every square inch of my epidermis was tingling. There's "Forest Bathing" in the trees and now there's "Epidermis Tingling" among the wings. As my good buddy Danny Barrett used to say, "If this doesn't turn your crank—it can't be turned!" Luckily for me, they didn't bite!

It was the ultimate **ASMR** high! You know—that "**A**utonomous **S**ensory **M**eridian **R**esponse" which causes a tingling sensation in your neck and head after triggers like people sweeping, whispering, or being covered in thousands of insects? Getting your hair cut may produce the ultimate ASMR response in humans and can be compared to the calm that primates get when grooming each other to remove parasites. Most people describe the sensation as relaxing and pleasurable.

I'm not kidding— it's a real thing! Look it up if you don't believe me! But apparently, not all people get ASMR feelings, or if they do, don't recognize the subtle sensations. Scientists have only recently started studying ASMR, but emerging research suggests that it can lower heart rate while improving overall health and well-being. And no—it's not a sexual thing—at least for most people.

Perhaps in our not-too-distant past, what began as simple grooming for parasite removal led us to develop tighter bonds, resulting in more emotional and intimate relationships. Nearly all primates practice grooming for both parasite elimination and emotional bonding, so are likely experiencing ASMR much in the same way we do.

*I guess I was just reverting back to my inner primate!*

The snow monkeys of Japan spend much of their free time in natural hot springs soaking up the warmth. Who can blame them—right? They also expend a great deal of time picking ticks and fleas off each other.

In fact, they enjoy the grooming so much that researchers have observed certain individuals picking ticks off nearby deer to place on themselves for the singular purpose of stealing some extra ASMR time. The deer also like the parasite grooming as well—an ever-developing symbiosis between two different species.

It would appear that our closest relatives are digging the good vibes touching one another brings. Perhaps ASMR and its calming effects have been around for as long as primates—nature's way of soothing that primordial beast that resides in us all.

That gives me an idea. Wonder if I could package up floodflies and sell them as **"Floodfly Ecstasy"** an all-natural, non-toxic, and sustainable rain forest product that is also completely biodegradable? The label might read,

*"We guarantee the ultimate ASMR jungle high all without any known or harmful side effects except for the occasional and accidental ingestion of a few bugs.*

**Warning:** Overuse of this product has been shown to cause mild cases of ASMR Dependency."

# Bikes, Buses, and Boats

You may have seen the movie "Planes, Trains and Automobiles" with actors John Candy and Steve Martin. This tale is not a make-believe movie but the reality of what happened to us after three weeks in the jungle—eating rice and beans three times a day, seven days a week. Don't get me wrong the food at BFREE is well-prepared and about as tasty as rice and beans get but there's only so much a palate can take.

It wasn't cabin fever we suffered from it was food fever and zeal for something other than you know rice and beans. Maybe if I whispered, the manna from heaven would just quietly and simply go away. No such luck, however. The bowl of "Dueling Palate Killers" would be patiently waiting for our return.

Living in the jungle with painful scorpion stings or lack of solar power doesn't hold a candle to the mouth-numbing chow and deep food depression that follows weeks of the same old thing. You begin to develop food scabies of the tongue and there's only one cure—a diet rich in fried snapper and smashed (not mashed potatoes) at Omar's Creole Grub with a Tutti Frutti gelato chaser or an almost Greek salad at the Shak.

After all, we were spoiled Americans when it came to food, raised on a wide-ranging diet that included every kind of food at our fingertips. But in Belize, what you see is all they have. If it's not in season, you don't have it. Still, the food away from BFREE would bring excitement to the 10,000 taste bud-receptors throughout our oral epithelium.

After weeks of the same old rice and bean meals, our snobbish food getaway would always be a weekend at Placencia, a small friendly town built on white-sand beaches and aqua blue waters. There we would enjoy the best darn food this end of the world. But it wasn't an easy trek, taking most of the day to reach the culinary bliss even though it was a mere 25 miles from BFREE. First, we had to travel by bike, bus, taxi, and finally, by boat to reach this food oasis.

**Bikes**

Our trek to food bliss began in early morning hours as we waded knee-deep across the beautiful Bladen River with our bikes in hand, as visions of snapper swam in our heads. For the next six miles, we would ride, carry and drag our bikes along the road of lost soles to the highway where we would hopefully catch the bus in time. Why didn't we just ride our bikes the whole way? The status of the road at any given moment was either dry enough to drive a school bus down, or during the rainy season, thick sticky mud that gummed up your wheels and chain until your bike weighed more than you did. Then it became more of a tug of war between you and the bike. The bike usually won so you just cuss and toss it to the side of the road and start walking. The bikes are great when the road is dry and takes around thirty minutes while walking the six-mile road through a scorching, shade-less savannah in under 3 hours is considered a pretty good pace. But it doesn't come without the heat and blisters of foot travel.

**Buses**

Our journey to food ecstasy moved to the next mode of transport—a bus. It was not an air-conditioned Greyhound bus with comfortable seats and plenty of legroom but a broken-down, overcrowded old school bus from the USA. And when I say overcrowded—I mean bursting at the seams with ample-asses in your face, unpleasant, nose-burning, cologne-covered people, and their shit stuffed to the ceiling. There was usually standing room only and when that ran out, they stacked people sideways, longways—anyway they could.

As we raced down the road, the jam-packed bus swayed back and forth like a boat in stormy weather. The bus driver drove like a bat out of hell, risking life and limb as we sped to the next destination—the small coastal town of Independence, but not before stopping for everyone and their goats along the way.

**Taxis**

At the bus station, we hailed a banged-up taxi, driven by an overly coifed and cologned driver for the 5-minute ride to the tiny village of Mango Creek situated on the Placencia Lagoon. This mode of transport was uneventful unless you considered the humorous sight of me hanging my head out the passenger window of the taxi like a dog to avoid the ravages of the artificial scent-laden driver. We continued on for a short ride or walk to the booming town of Mango Creek on the Placencia Lagoon, a village so small you might miss it if you blink. There we hopped on our next mode of transport—a boat to Placencia.

**Boats**

Finally, the last and final leg-weary, mental-siphoning part of the trip—the Hokey-Pokey Boat Dock and local mango tree of gossip. It sits next to the rickety outhouse built over the lagoon which fed the fish waiting for the next mouth-watering human excrement droppings. Yummy!! The boat ride began in a water taxi that held approximately 15 people with their gear and anything else they could fit in the boat. Just like the bus, it was jam-packed to near tipping.

The boat captain, an ample young man of considerable girth, pushed the throttle down to full speed ahead as our butt-cheeks slapped the rock hard fiberglass seats, bouncing over teeth-chattering waves for the next 20 minutes.

As is common in Belize, about 10 minutes into our ride, the rain began to fall in sheets. With only a Bimini top to shield us from the onslaught, out came plastic tarps to shed off the downpour as each passenger grabbed a corner and held on to the tarp for dear life.

As we rocketed past mangrove-bordered estuaries teeming with Red-footed Booby's and Magnificent Frigatebirds, the last ride of the day would result in soon-to-be culinary delights of very imaginable enjoyment.

After enduring a day of rice and beans, mud-covered bikes, blistered soles, a bus, and taxi that reeked of street whores, and last but not least a butt-numbing, brain-scrambling boat ride—at last, our destination of Placencia was in sight.

But it was all worth it when those first bites of redemption passed through those pie-holes. Our mucus membranes were titillated not with rice and beans but with lobster grilled to perfection on a bed of wood coals. It's the kind of food that makes being human and having consciousness so damn rewarding. It's the capacity to fully appreciate life's simple pleasures.

We would come to repeat this scenario again and again throughout our seven years *living where the cats eat the dogs.* These monthly jaunts provided a reprieve from the challenges of living in such a remote place. Like so many other things, we no longer take food diversity for granted and try to keep a healthy perspective on how and where our food comes from—even that lackluster bowl of rice and beans.

## I'll Never Wash My Butt Again

You may be familiar with the program "Doctors Without Borders", a group of volunteer physicians who donate their healing skills in remote areas without adequate access to health care? But did you know there are pilots who donate their time and skills to provide much-needed conservation flights to protect natural resources in other countries?

Every year these part heroes, part crazies from around the world risk life and limb to fly politicians, land managers, scientists, and biologists over remote regions of Belize providing a bird's eye view of the surrounding landscape. These flights are conducted to gather information, pinpoint illegal logging operations, deforestation, and forest fire damage from slash and burn farming practices, and even locate magnificent Ceiba trees that are potential nest sites for Harpy eagles.

You're no doubt wondering about the life and limb comments? Well, it's quite simple really—it's my extreme fear of flying and my life and limb I most worry about. For me, being trapped inside a tin can with wings feels like bungee jumping without the bungee. I have caught crocodiles and deadly snakes, come face to face with jaguars, barely escaped a helicopter crash in Montana, almost burned to death in a California wildfire, very nearly died from the bite of a serpent, and narrowly cheated certain death from a 1000-foot-deep crevasse I fell into on Mount Shasta. While each of these life-threatening events were scary enough, for me, none hold a candle to the fear I feel in an airplane. It's clearly my Achilles heel. Once airborne, I go from untamed and fearless to a terrified, wimpy little puppy. When my feet leave the ground, I struggle with FFA or Frightful-Flight Anxiety.

I know it's not rational anymore than most baseless fears of snakes and spiders. But I like to think that at least some of my fear is grounded (where I prefer to stay) in reality. While there are several decent airports in Belize, most are little more than dirt runways. Furthermore, in a small Cessna that is not instrument-rated, there's no radar warning of other aircraft in the air so the pilot must use Visual Flight Rules (VFR). Nearly year-round cloud cover from the steamy moisture of a tropical region often creates conditions for poor visibility—and once airborne over an endless sea of jungle in mostly mountainous terrain, there's nowhere to safely land.

If you go down, there's only "Splatsville", although Alan Rabinowitz and his pilot did manage to survive a plane crash while conducting aerial surveys for jaguars in the 1990s in the Cockscomb Basin Wildlife Sanctuary. The small plane hurtled downward into a tangle of vines and trees which likely saved their asses. My worst nightmare!

*One time at band camp*, no that's a movie, what I meant to say was one time during a flight over the Maya Mountains with one of those "pilots without borders" organizations, Lighthawk, our plane became hopelessly lost in an endless sea of fog. Dense cloud cover had moved in, causing a total "whiteout" or you can't see shit out there!

We usually flew with the doors off the plane so we could see and photograph landscapes better. It's an eerie sensation looking out and down seeing nothing but unyielding fog and no sign of land. Of course, you're strapped in so you can't accidentally fall out. But it's still an unnerving, gut-wrenching feeling—a mental arm-wrestling match between logic and fear with your brain shouting—**I am going to die**. Unable to distinguish mountains tops anymore, the pilot turned to me and asked "What elevation are the highest peaks here?"

At this point I wanted to shout, "Why don't YOU already know the answer to this?? Did you forget to do your homework?" But reason took hold. After all, this guy literally had my life in his hands. No time to be a smartass. *"Around 3800 feet," I stammered.*

"We need to be at least several hundred above them to avoid a collision," he said, which would certainly lead to an unhappy ending. Feeling nauseated and suffering the cruel effects of FFA, I was beginning to sweat not bullets but rice and beans eaten just moments before takeoff. The only thing worse than rice and beans going down are those same rice and beans coming back up. It's like when you get too close to a turkey vulture that responds by regurgitating its maggot-filled, road-kill opossum dinner in your face. It wasn't at all a pleasant thought, but somehow I managed to keep the rice and beans downstairs where they belonged. It would be just my luck to die in a plane crash and rice and beans would be my last meal and demented memory. It would serve me right for poking so much fun at you know—rice and beans.

With no clear way out of the murkiness, the pilot was essentially flying in random circles, or so it seemed. And just when I thought it could not get any worse, a little light on the control panel of the plane came on.

*"Hey, what's that light that just came on?" I shouted, forgetting that I was talking into a headset.*

"Well, it's not good," he replied. "The first fuel tank just ran out. But don't worry, we have an emergency tank."

DON'T WORRY?? YOU HAVE GOT TO BE KIDDING ME???

Now on the second, "emergency tank", we were running out of time to land our butts on the ground where they belonged.

You can only imagine where my head was right about now. You guessed it right—between my legs, kissing my bony ass goodbye! I would rather have been in a cage filled with spitting cobras than in the soon-to-be pile of scrap metal and disfigured bodies. Judy would have little to bury and worse yet, no organs to sell on eBay since I'm an organ donor.

As it turned out, the pilot eventually squeezed the plane through a small hole in the fog and, after seeing flat land, made an emergency landing with no time to spare, the second tank nearing empty. I wondered was it **Itzamna**, Sky God of the Maya that saved our worthless rears, or just plain crazy damn luck?

Nevertheless, I would push through my fright of flights for the scientific gains. During the next 7 years, I would make many more flyover trips throughout the Maya Mountains, locating new research stations to study you guessed it, land snails. But it never really got easier.

All the pilots that volunteer these flights are of the highest quality, volunteering time and money in the name of conservation but did I tell you that some pilots are just plain crazy? I would have the pleasure and terror of flying with one such pilot, Bruce Gordon of Ecoflight.

The first and most memorable flight with Bruce would be one of sheer panic! I was first introduced to Bruce when I scored the co-pilot seat on one of our conservation flights. During a bumpy takeoff from one of those pseudo-runways, with a tummy NOT full of rice and beans, we began to ascend. As we became airborne, Bruce asked me to be on the lookout for unexpected flying machines to steer clear of any midair collisions.

It wasn't very reassuring, but by now I was somewhat used to the drill. Leaving the savanna behind, we began to encounter the usual misty weather but continued onward and upward toward the Mountains of the Maya. In the distance, we could see the rocky crags of the Cockscomb Range, one of Belize's highest mountains and most celebrated landmarks.

As we got closer to the plane-crushing rock, I noticed that Bruce was reading the freaking flight manual instead of gripping the steering wheel or whatever you call that thing his hands should have been glued to!

*I screamed into the headset, "Bruce, we're headed straight for that cliff face. Shouldn't you do something like—let me think, steering the fucking plane?!?!?"*

It was like he was in la-la land, paying little attention to me, and no attention to the inevitable collision course we were on with certain death as the inevitable outcome. Just seconds from becoming a red stain on the Cockscomb Peaks, he grabbed the wheel and banked the plane to the left. We were safe.

*"Ha-ha very funny, you do this to all your newbie's don't you?" I accused.*

There was little more than a pathetic snicker from Bruce but for me a new pair of shorts and 6 months of therapy. I decided I should try to get to know this crazy, jokester who was ferrying me around in the sky that day, so I innocently asked,

*"Where are you from, Bruce?"*

"Aspen, Colorado" he replied. Now, while I didn't personally know anyone from there, I certainly knew that my favorite singer of all time lived there before his untimely death in—guess what?—a plane crash!!

When I mentioned that John Denver, my favorite singer, lived there at one time, Bruce's eyes widened. "John was one of my best friends! We often flew together in Colorado," Bruce proclaimed. "In fact, it just so happens that John would sit in that very same seat you're sitting in right now!"

*"Wow! Really? I can't believe it! I'm sitting in the same seat that John Denver sat in? How cool is that?" I said like the John Denver fanatic that I am.* ***"I'll never wash my butt again!"***

This proclamation sent Bruce into a fit of laughter and began a friendship and camaraderie. We were kindred jokester spirits.

After safely landing, we posed for pictures in front of the plane that John Denver had flown in. My heart was full as Bruce and I belted out the song, "Take Me Home, Country Roads" in honor of his friend and my idol.

Little did we know when we began this adventure, *living where the cats eat the dogs*, the extraordinary people we would meet and befriend. Bruce has received many awards including the prestigious "Conservation Hero Award" and has flown over 10,000 hours of conservation missions in multiple countries, mostly in bush and mountainous conditions.

And so goes the story of the unwashed buttocks which actually lasted only a few hours in the hot and sweaty jungles of Belize.

# Mayan Weddings and Sex Meat

Maya weddings are part Catholic, part Maya but mostly just old-fashioned, raucous fun. Judy and I had the honor of being asked to take part in several Mayan weddings while in Belize. Many Belizean couples forego the trappings of marriage, however, to just live together without ever participating in the silly-religious or civil ceremonies.

Nevertheless, we were both excited to celebrate the bonding ritual of Carolina Cus, our adopted Mayan daughter, and most amazing BFREE cook. The day began before the sun dared show its face in the darkness before dawn. As we climbed into Grizzly for another 6-mile slog through the mud-covered road of lost soles, Emily and Kevin, avian researchers from Canada, joined us for the 3-hour drive north for Carolina's wedding.

Our first Mayan wedding experience had been the year before in the tiny village of San Miguel, but now we were headed to the not-so-big capital city of Belmopan. For me, the wedding ceremony was like most; I couldn't wait to get through the lines of bullshit to get to the lines of food, as long as it didn't include rice and beans.

The preacher, a man of Kriol not Mayan descent, was like a car salesman trying to sell you that car you're not the least bit interested in. He went on and on for what seemed like days. Boredom was beginning to weigh heavy on my eyelids. I was slowly slipping into unconsciousness as the preacher continued on about love and commitment and why lust was a sin and why we should be true to only two things—God and marriage.

While most everyone around me was digging it, I was thinking I'd rather be in a flying tin can over the Maya Mountains about to run out of gas, which I wished he would. Up until this point, the preacher was about as interesting as a five-pound sack of fertilizer.

Just as I was ready to grab the microphone and start singing "We are the World", the preacher uttered these uncharacteristic words.

*"Husbands don't treat your wives like sex meat".*

My eyes popped and my brain now throbbing with a renewed interest, I glanced over to Judy and whispered, "Did the preacher say what I think he said?"

With a startled and comical look on her face, she nodded. I cast a sideways glance at Kevin and Emily whose shoulders were shaking—not from a desire to speak in tongues—but to prevent an outburst of laughter! But they dared not, for fear of suffering the all-mighty Maya God of smite. Like the dutiful evangelical preacher he was, he repeated his best line once again, just for good measure.

*"Don't treat your wives like sex meat*
*or lustful objects*
*but with respect and dignity."*

In a matter of seconds, this man had just gone from that "annoying car salesman" to "preacher of the year" in terms of rousing my coma of boredom.

He continued on—not about marriage, but about lusting and cheating on your wife and what sounded like his many inadequacies and short-comings as a man. It was like being on an episode of "True Confessions of a Preacher".

What the fuck was he talking about? The message was clear as mud as they say in Kentucky. But we did manage to get through the rest of the ceremony with our sheepish grins and bottled-up laughter to honor our dear Carolina. Or so we thought.

The quirky preacher would deliver one more surprise at his redemption, I meant Carolina's reception.

My wife, who has a beautiful singing voice, chose to honor Carolina with a song at her raucous reception. Flanked by giant speakers taller than her, she walked up to the microphone and began her tribute.

♪*You fill up my senses....like a night in the forest.*♪

The song should have been "you're a terrible leader…because you're a nightly cheater".

Nevertheless, Judy's voice rang through the reception like a Solitaire in the mountain mist as Carolina and her beloved smiled in acknowledg-ment of the gift.

But wait—what was that noise in the background?

A slightly off-key male voice was intruding the lovely rendition.

**He's Back**! The preacher that is. He strolled up to the microphone, by this time somewhat inebriated, and chimed in.

Judy, to her surprise, stumbled on a few words before recovering to fin-ish the song as an unlikely duet with the wacky pastor. Nonetheless, we were there to support our good friend Carolina, not for the ranting of a madcap preacher. Although I will have to admit—it was by far the most attention-grabbing wedding I had ever attended.

# Lost River Cave

Deep within the jungles of the Bladen Nature Reserve flows an extraordinary waterway living a binary life; one open to the sky and mountains beyond and, the other of unforeseen secrecy. Little of it terrestrial, most of it subterranean, this lost river courses through and beneath pristine rainforest. It's a journey born in darkness and mystery where great caverns are filled with weeping rocks and fantastic creatures. In the truest sense, it is a lost world that has never been seen. For the eyeless creatures that live there, it remains invisible in a realm of perpetual darkness. This is where the river begins one drop at a time. For a millennium this place has held secrets beyond human reach locked away in an underground world.

For decades this uncharted river was thought to be a split of the Bladen River system. But in 2014, Judy and I decided to walk the jungle river to its beginning. What we would discover wasn't a split in the Bladen River at all, but a completely new river emerging from deep within the Maya Mountains.

Bushwhacking through thick lianas with no visible trail except the beaten path that tapirs and jaguar traversed, most of our journey would be spent wading in the Bladen River up to our waists, at times, hiking packs hoisted above our heads.

While I had spent many days and nights on expeditions deep into the interior of the Bladen Nature Reserve—Belize's highest protected area—conducting snail research and assisting the Avian Team, Judy was usually busy with the management of BFREE. The trip would be her first expedition into this pristine, almost uncharted area.

After three hours of trekking, we reached the part where the Bladen River split. Continuing our slog through the water, we turned left and headed up the narrowing stream. Deeper and deeper into the unexplored territory we crept until the waters disappeared. We could go no farther.

"But wait, there's a small opening, it looks like a cave," I said. "I think we can squeeze through it with a little room to spare."

As we went deeper into the abyss, it began to open in size. The next few minutes would become Judy's worst nightmare.

"Hey, Judy, look at the large colony of jumbo bats over there hanging on the wall!" I said with the excitement of a seasoned bat biologist. Startled by our presence, the bats exploded into flight.

For me, the sound of large beating wings in a hollow cave was nothing less than astounding.

For Judy, however, it was the Alfred Hitchcock movie "The Birds" coming true. As a kid, this movie would set into motion her lifelong fear of flying things. I glanced back to see my usually adventurous and somewhat fearless wife, plastered up against the cave wall with a look of horror on her face.

*"I can't do this anymore! I am outta here! I will wait at the entrance."*

I absolutely understood her fears but having worked so hard to get here, and with curiosity about the real origins of this mystery waterway, I would push onward to explore the remainder of the cavern.

Large bats were flying everywhere as I swam against the ever-stronger current.

As the passageway got smaller, so did the space the bats had to miss my face and tangle in my hair to lay eggs. Just kidding about the latter, bats don't lay eggs.

I was now in water over my head as bat wings wisped past my bobbing cranium.

"Does it get any better than this?" I thought. Not for a habitually wacky, rogue biologist! But the cave had run out of dry passageway and now my entire body except my bobbing head was underwater, so I could go no farther without a dive suit and tanks.

Our exploration debunked the theory that the stream was just a split of the Bladen River, but in fact, an unnamed river flowing out of the depths of the Maya Mountains, a Lost River in time and place.

It was time for me to turn back and leave the underworld of the dead, **Xibalba,** and return to the surface of the living. There Judy would be waiting as always, with patience; a bit shook but no less the tough lady that she is.

*What's that in your hair?*

# That Sixth Sense

There are those precious few moments in the forest when your movements forward are interrupted by something that—at first—goes unnoticed. There's a pause in your stride, a calm in your breath. All five senses now amplify, filling to capacity with endless waves of incoming mental and physical stimuli. What did I fail to notice???

It's like you're frozen in time—searching and reaching for that "instinctive awareness" that's gone missing. It's when **Sight, Sound, Smell, Taste,** and **Touch** reach a kind of equilibrium, bringing forth that rarely experienced "sixth sense", although many would probably describe it as something more mystical. Whenever I lead walks in the woods, I encourage people to engage all their senses. We don't just look at a plant but also touch it, smell it, listen to the sound the leaves make in the wind, and at times, taste it. This way all the senses are drawn into, and become part of, the experience.

As a biologist who spends a good deal of my time in the woods—paying closer attention to my five senses has not only helped me better understand nature—but also kept me safe from nature. Like being cognizant of the slightest sound a snake makes crawling over dry leaves or how the buzz of a hornet is vastly different than the buzz of a hoverfly. Paying close attention to these subtleties has kept me from stepping on venomous snakes or walking into a hornet's nest. By taking the hurry out of time and allowing our five senses to coalesce, our surroundings become more finely tuned. It's like that favorite radio station that needs a slight more tweaking to hear that song a little clearer.

Earlier humans no doubt had a better handle on their sixth sense than we do today, ours largely dazed and confused from spending too much time inside little boxes of wood and plaster—a synthetic world of sensory-killing lifestyles unlike any of our ancestors could ever imagine. The sixth sense no doubt kept our distant ancestors alive and would serve us well on our next jungle expedition below.

Judy and I decided to go on a canoe trip up the Bladen River for an overnight camping trip. You heard me right we went up the river not down. Remember I'm dyslexic and so I do everything backwards.

We would paddle the pools and drag through the ripples for hours and miles until nightfall when we would set up camp. Reaching Holey Rock (named for its many holes) by noon was the halfway point and a welcomed lunch break. The rock feature is a favorite swimming hole for jungle weary researchers seeking a little thrill, jumping from the many cavities above the cool clear waters of the Bladen River. With bellies full of peanut butter and jelly sandwiches—a biologist's best friend next to a Ziploc bag that doubles as my wallet, and when necessary, a place to keep snails—we were ready to push onward. Finally at camp, we began to set up our jungle hammocks. With both heads buried in our packs retrieving some camp gear, I felt the presence of something close by—the hairs on my arm and neck standing at attention—that sixth sense was now kicking in.

"**Don't Move!**" I whispered to Judy who's butt was facing the unknown intruder.

Someone or something was close by. I slowly reached for my camera when I recognized our unexpected visitor. Judy, however, had no idea who or what was invading our camping site! Was it a Fer-de-Lance or a hungry jaguar looking for an easy meal? I managed to get a few shots off and then quietly told Judy to slowly turn around. The creature walked right past us down to the river for a drink. Judy turned just in time to catch the animal walking back toward us. Still unaware of our presence, the creature stopped and sniffed the air with its prehensile snout. Now, just meters away, yet oblivious to our presence, the mountain cow slowly disappeared into the bush.

This legendary creature known as the Mountain Cow in Belize and Central American Tapir elsewhere is a large, semi-aquatic mammal weighing as much as 600 pounds. Related to the rhinoceros and although rare elsewhere, the tapir is a common resident of the Maya Mountains and Bladen Nature Reserve. Equally at home in water or on land, the vegetarian tapir has poor eyesight, but a keen sense of smell. Its large three-toed tracks are a common sight here in the Bladen. It was indeed one of our most memorable wildlife encounters together while at BFREE.

In steep contrast to the tapir adventure, there are those treasured moments in the jungle when your five senses are besieged by something quite foul. It's the *Sights* and *Sounds* of bouncing tree branches from above, the barnyard *Smell* of falling poo, the *Taste* of urine raining down on your face and a feeling that you have been *Touched* by **Hun-Chowen**, Monkey God of the Maya, or in simple terms, a spider monkey's way of saying hello! If these kinds of monkeyshines don't turn your crank, then it's probably in need of some WD-40.

# Cyclops the Tapir

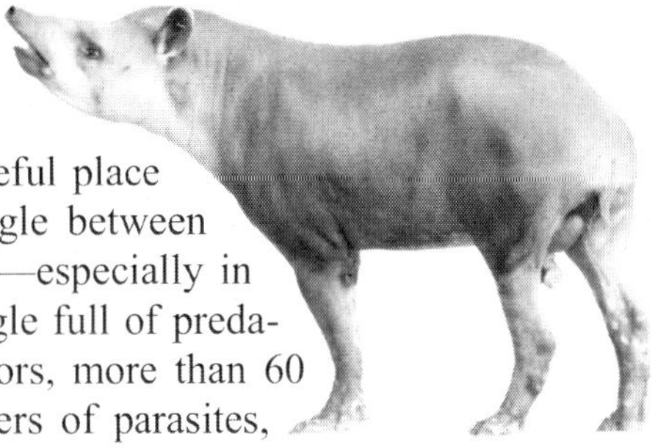

The jungle is not the calm and peaceful place you think it is. It's a constant struggle between life and death—*a cat-eat-dog world*—especially in Belize. Surviving attacks from a jungle full of predators including 5 native cats, 28 raptors, more than 60 species of snakes, and untold numbers of parasites, all seeking to turn you into a meal or breeding ground, can be a daunting task for wildlife. An animal's best defense is simple. Stay concealed as long as possible until it can't. When it needs to feed, do it quietly and stay alert. Even if it follows these basic rules, it will still be hunted, caught, killed, and consumed. It's the law of the jungle. But just because an animal is hunted and caught, doesn't mean it's always killed and eaten. Many such instances turn out in favor of the hunted, at least in terms of a successful escape, but usually notwithstanding some lingering damage to flesh and bone. The challenge then becomes surviving the next few weeks until you heal. Then there are the emotional scars of nearly being ripped apart that perhaps never entirely go away. Sounds a lot like our own lives doesn't it?

On the other hand, it can end badly for the hunter. A broken or torn leg muscle from the tusk of a warrie (wild pig) can prevent a jaguar from hunting—leading to starvation, while a broken wing of a raptor from a tangle with its prey is certain death. We saw countless animals with limps, scars, and missing parts from their near-death experiences. While canoeing up the crystal clear waters of the Bladen River one evening we saw a Morelet Crocodile that had inch-deep claw marks down its back that had healed. This was no doubt the work of a large cat that lost its grip on the seven-foot crocodile as it dove into the river.

The "balance of nature" is an outdated theory that puts forward the idea that ecological systems are usually in a stable equilibrium or homeostasis, but this is a myth. The theory has been largely discredited by scientists working in ecology, as it has been found that constant disturbances (e.g. storms or fires) leading to chaos and energetic changes are the norm rather than a perceived balance. So the "**the balance of nature**" could be restated as "**Natures' Sway**", swinging back and forth like a pendulum, over and under population from a constantly wobbling ecosystem and the resources required to keep it all running more or less successfully. When a species swings too far one way, the "laws of nature" self-correct by decreasing or increasing predation, cannibalism, parasitism, disease, starvation, and/or homosexuality. These factors, either alone or together, keep the population in check.

If, for example, there are too many bunnies hopping around a field, nature may reflexively respond by increasing predators such as foxes, as one of several control mechanisms. This in turn knocks the bunnies back down to a sustainable level or even below—the “pendulum effect”. Land has a "carrying capacity" for every organism including humans, and if exceeded, will sooner or later seek out impartial stability, sometimes with deadly consequences.

While most species in their native habitat have no choice but to live within these constraints, ***Homo siphons*** have circumvented these basic imperatives by means of ever-evolving brains and the planet-destroying technology that follows. This has allowed us to jump through and ignore these self -governing systems. But we will only escape the wrath of nature or our self -destructive behavior for so long before one or the other comes back and bites us in our ass, or in the case of a virus—our immune system!

One animal that fell victim to an attack but survived was Cyclops, the name we gave a tapir with a missing eye and a shredded backside of scars—no doubt from the attack of a hungry jaguar. For weeks, the emaciated ungulate was often seen at the river crossing and become somewhat accustomed to our presence. One day intern Katelyn was working on the road of lost soles when she heard a loud commotion coming straight toward her—Cyclops, the tapir, with Jelly the dog, in hot pursuit. The tapir was running straight for Katelyn, unable to see her due to its blindside. Hoping to avoid a head-on collision, Katelyn leaped to safety just in time—avoiding being plowed into the ground by the 500-pound beast. Fortunately for Kate, the road of lost soles would save her that day.

We would see Cyclops, the tapir, for a few more times and then she vanished. Likely, the creature had succumbed to its injuries or became dinner for a large cat. But this is the way of the jungle; that circle of life that feeds on death and bequeaths back life.

# Woo-frogs, Psychedelic Toads, and Spicy Anurans

## Woo-Frogs

After the epidermis tingling of the floodflies emerging at the beginning of the rainy season in late May, it was time for the auditory enchantment of Woo-frogs. The Mexican burrowing toad or Woo-frog as they are known in Belize, begin to stir in the ground about the same time as the floodflies. It’s been a year since they have seen the dark of night.

They have been waiting patiently for the heavy rains for one reason—a chance to amplex, or in layman's terms, proliferation. But for frogs, the whole event is done externally so for the act of mating is by no means the same fun we have!

Technically, amplexing is the act where usually smaller males ride on top of the females and fertilize eggs as they are expelled. That's it in a nutshell! For both males and females, it is less than a climactic finish to a long-awaited event. Nevertheless, it gets the job done.

It all begins when the usually motionless toads emerge from their earthly tombs of boredom. The collective horde, now free from the shackles of lethargy, burst into a breeding frenzy like a whorehouse during Mardi Gras. Untold numbers of deprived and horny males emerge at once and began to call in earnest for females as the night air becomes heavy with lust and anticipation of a score.

The males make the woo-calls from shadowy realms where they swell with self-importance, like over-anxious adolescents coming for the first time. The females are not attracted to the biggest or handsomest males, but the ones with the most to bequeath. And as soon as the orgy-fest begins, it's all over in a matter of days. Little remains of the sex party of the year except viscous condoms lining the streets of New Orleans—I meant—gelatinous eggs laid on the bottom of scattered ponds.

The Mexican burrowing toad is genetically unique in a number of ways. It's the only species within the only genus of the family Rhinophrynidae. Woo-frogs have had over 190 million years of independent evolution and are one of the most evolutionarily distinct amphibian species on Earth today. Believe it or not, a fruit bat, polar bear, killer whale, kangaroo, and human are all more similar to one another than Woo-frogs are to any other amphibian!

## Psychedelic Toads

The infamous marine or cane toad, *Bufo marinus,* is the largest anuran (size of a small dinner plate) found in Belize and the world's largest toad. The species is well-endowed with hefty parotoid glands that contain copious amounts of whitish poison bufotoxin (an alkaline-based poison) usually fatal to anything that attempts to include it as a meal.

Marine toads are native to Belize, so most animals there have learned not to eat this generously proportioned and highly toxic toad and their eggs or tadpoles. Natural predators of cane toads, unaffected by the bufotoxin, include cat-eyed snakes and several species of catfish. Unfortunately, marine toads have been introduced to a number of countries like Australia and the US to control cane beetle pests where they have become the dominating nuisance, wreaking havoc on native wildlife populations.

Ingested in small quantities, marine toad poison produces hallucinating effects and was used by the ancient Maya in ceremonies and spiritual rituals. Present-day toad-lickers in the US prefer the less toxic Colorado River Toad for their psychedelic journey. According to most toad-lickers that are still in their right minds, one of the unique effects is tactile enhancement, which can be an overwhelming full-body sensation, especially near the nerve endings. Sounds like the feeling I got from epidermis-tingling floodflies but without the dangers of toad toxin which has killed a number of inexperienced people. Cane toads breed and lay eggs in pools found in medium to large river systems throughout Belize. They can be found in shallow burrows along steep banks of creeks and rivers year-round.

### Spicy Anurans

The pepper treefrog is a large tree-dwelling anuran adapted to live in a variety of dissimilar environments including savannas, evergreen jungles, and dry forests, from sea level to over 6000 feet. The large vocal sacs used to make strident calls are also employed as a life preserver, keeping the males afloat while calling for females. These treefrogs are capable of gliding maneuvers as well, moving them safely from treetop to jungle floor without little harm. Most importantly, these frogs are decidedly toxic and therefore, should never be handled without the use of plastic gloves.

If disturbed, the frog can manufacture copious amounts of a latex-like poison used to protect against predators. It is sticky, water-insoluble, and not easily removed from skin. According to reports from Belize, the pepper treefrog's poison is said to cause sneezing, hence its common name. I know from personal experience that wet or dry, the poison causes severe burning, likened to hot pepper juice, if accidentally rubbed into one's eye, smarting for days.

## Don't Worry, It's Only Warries

For anyone who has spent much time in the jungle, wildlife encounters are undoubtedly the highlight of nearly all trips. It's an ecosystem on steroids where every nook and cranny is jam-packed with life. From crawling land snails to great herds of the endangered white-lipped peccary locally referred to as warrie, it's a place of plenty. While most wildlife experiences are enjoyable, some can be downright chilling to the bone. Take warries for example. These swine are legendary, both in herd size and fierceness, toward impending threats such as jaguars. They use scent glands to mark territory, so they are more often smelled than seen. It is among the most feared wild animals in Belize and the Americas. When threatened, warries most often run but will also stand their ground, facing adversaries while making ear-piercing clicking sounds with their frightening canines and have even been known to attack humans.

Watching these pigs fight is like watching two chainsaws battle for dominance. Teeth slashing, bones breaking, blood and flesh flying everywhere. It's a blood bath. In large numbers, this species of peccary is a force to reckon with!

I had one such experience that would scare the crap out of me but was also strangely gratifying. It was one of those rare sunny days in the rainforest. The Avian Team and I were on one of our monthly research treks into the jungle, when a foul odor dominated the jungle. "What's that smell and clamor coming down the hill?" I asked William.

*"It's the devil himself, coming to collect our souls or worse," he teased.*

It was in fact a massive herd of warrie coming straight for us!

*Oh shit! What do we do?*

Nothing to do but hope and pray to **Cit Bolon Tum,** the Mayan boar-headed god of medicine and healing. Surely he or she wouldn't send the swine to dice and slice us to pieces just to then heal us? But just in case Cit Bolon Tum wasn't listening that day, we decided to hunker down behind some rather meager bushes. As our bodies trembled with fear, the herd had finally reached our location—yet appeared unaware of our presence, at least for now. As they began to pass us by, we thought we were done for. There was only warrie cohune palms to climb—armored with thick slashing thorns—hence their name. Would we be torn to pieces by the warrie pigs or by the warrie trees? There was only one thing to do at this point, kiss our asses goodbye and hope for the best.

They were passing us on one side first, and then on both sides. We were now in the middle of the herd of more than 200-strong but to our surprise, the herd seemed oblivious to our presence. We didn't dare to breathe or move a muscle.

But then one of the males stopped, stuck his nose up in the air, and said—ok maybe thought—"Fee-fi-fo-fum I smell the fear of two feeble *Homo sapiens.*"

He—the pig that is—began clacking his tusks, followed by the others until the sound became overwhelmingly deafening, not to mention the eye and nose-burning stench. All I could think of was bone-crushing jaws and tearing flesh, leaving nothing left for Judy to bury. But that sun-drenched day the warries picked flight over fight, leaving us with a bitter/sweet taste of jungle impartiality. In a hasty and disorderly stampede, they were all gone within just minutes of arriving. It could have turned out very differently that day!

## Wrestling with Fairytales

Looking like something from a fairytale, anteaters are extraordinarily evolved, toothless mammals that have twelve-inch-long, worm-like tongues used to harvest termites, ants, and occasionally, bees. These uncanny creatures reside in trees, liana-tangles, and bridging branches; using their massive front claws to rip open tough termite hives accessing the protein-rich insects within. The claws can also be employed as lethal weapons, using the enlarged middle claw to inflect deep lacerations in the toughest hide. The domestic cat-sized tamandua is the most common anteater in Belize. The dog-sized giant anteaters of South America (which have 6-inch long front claws) have been known to disembowel a jaguar in self-defense.

The much smaller, silky anteater is strictly arboreal; traveling on pencil-thin branches in search of its diet of ants. Using a single enlarged claw on the front feet, hollow ant-filled branches are easily sliced open exposing ants and their eggs. Out comes a five-inch long sticky tongue that laps up the fleeing ants.

My most memorable encounter with a tamandua occurred on one of my late-night excursions in the jungle. In the glare of my headlamp, a tamandua walking along the trail caught my attention. I hurried up to the animal in order to take photos and at first, it seemed indifferent to my presence, that was until I reached down and grabbed it by the prehensile tail.

That's when the tamandua broke loose of my grip and went berserk heading for the nearest tree. But, determined to photograph the normally elusive creature, I wasn't going to let it get away, so I chased it until I caught up with it again.

Why? Because that's what biologists do. We catch things, even if it means being bitten, poked, or stung. So I grabbed it by the tail but this time hanging on even tighter. By this time, the tamandua had climbed 4 feet up the tree. Holding its tail with my left hand, I reached for my camera. I wasn't worried about the toothless mouth or 12 inch sticky tongue but those 3-inch flesh-tearing claws. As the tamandua pulled harder for its freedom, I held on even tighter. This would go on for what seemed like hours until my left arm exhausted from fatigue had to finally release the tongues of fury. The next day I could barely move my arm fatigued from the night before. But I had some decent images of that enchanted fable of the night.

**Tamandua**

# Brawling with Crocodiles

If wrestling with fairytales doesn't impress you, perhaps brawling with titanic out-of-control crocodile will. My first and most memorable experience with a Morelet's crocodile would be one for the records.

It all happened one late morning with Judy the Cutie and Jake the Snake on the road of lost soles. We were driving out of BFREE to the local store to pick up supplies for an upcoming university field course when I spotted a crocodile in one of the many muddy pools along the road. From the back of the jacked up Toyota Tacoma with no brakes, I pounded on the cab and yelled, "Jake, stop Grizzly! Crocodile!!"

I jumped out of the truck with the enthusiasm of a kid in a candy shop, wanting to seize the over-sized lizard for a closer look. Why? Because it's how I, as a biologist, learn anything new about the critters I work with and protect. Besides, if I lost a finger or two doing my thing, there would be a doctor waiting to do their thing.

As I ran up to the crocodile, it slowly disappeared under a thick blanket of muddy water. No longer visible, and sensing my reluctance to blindly grab the beast, Jake said to me;

*"Grow some balls and just reach down and pull the croc out".*

I had caught lots of animals before but always had them in plain sight, where I could usually see the dangerous end. But this was different and Jake was egging me on. So what choice did I have? Freaking out about now and being the voice of reason was Judy, begging me to rethink what I was about to do. I had to choose between egging and begging.

My fool-hearted pride got the better of me, so I dove in. I was grabbing at anything that wiggled until I felt the hard Osteoderms or bony scutes on the crocodile's tail—the somewhat safer end of the prehistoric animal, although they have been known to use their tails to break bones. But at this point, a broken bone seemed better than missing fingers.

*"I've got it by the tail Jake and still have all my fingers!!" I exclaimed.*

As I pulled the creature from the waters, it was thrashing and biting like a chainsaw running at full throttle.

*"I'm not sure I can hang on to it Jake, it wants a piece of me, BAD."*

I finally pulled it from the muddy waters and we all were surprised at how big it WASN'T. If I had been a dung-beetle, it would have been enormous. As it turned out, it was less than a foot in length.

Judy, of course, was relieved while Jake cracked-up, laughing like a crazed hyena. But the last laugh was on him when I hurled the chopping, razor-sharp, teeth-filled mouth toward him, which he promptly caught by the mouth, you know, the dangerous end. Before he could turn loose of it however, the crocodile managed to inflict multiple lacerations using its razor-sharp teeth.

*"Who's laughing now, Snake Man?"*

I almost forgot. Why should the story be one for the Records—for having you fooled into thinking that this was some grand story, when it was nothing more than a lame-shame anecdote of microscopic proportions. I'm bad!

On another occasion, I was out one late night in the jungles around BFREE by myself when I came across a much bigger four-foot Morelet's crocodile in a small creek. Not thinking about safety first, I jumped in and quickly subdued the animal with my right hand and knee, while reaching into my pack for a camera with my left hand. It was like playing a game of Twister with myself, but instead of being on a plastic tarp, I was knee-deep in a creek with a crocodile in one hand and a camera in the other. It was a crazy balancing act, keeping the croc pinned down to keep from being sliced open while taking a series of photographs. It was a good thing Judy wasn't there that night to give me a hard time, which I most certainly deserved.

## Royal Rat & Bamboo Chicken

The most expensive meat in Belize is not the mouth-watering Caribbean lobster, grilled on a bed of natural wood coals, but a large rat or gibnut, as its known here in Belize. This essentially forest-dwelling rodent is limited to Central and South America. Gibnuts are one of the largest rodents on Earth, weighing as much as 18 pounds, only second in size to the largest rodent on Earth, the capybara, which can weigh between 60 to 170 pounds.

Unlike its smaller diurnal cousin the agouti, gibnuts are strictly nocturnal and seldom emerge until a few hours after sunset. When pursued by a jaguar, it may dive into ponds or rivers and submerge completely to escape danger. Like the agouti, gibnuts eat a variety of seeds, nuts, and greens. Because gibnuts are hunted for their prized meat, in some locations, the species has all but disappeared.

I would get my first taste of wild gibnut while on expedition into the Maya Mountains searching for Harpy Eagle nests. I would also meet Henry, the youngest of the newly-formed avian team for the first time. Henry, at age 17, was in many ways still a youngster, and struggled to make grown-up decisions. He had a childlike innocence about him, trapped in a young man's body and would receive more than his share of browbeating from the bird crew over the next few years. They were downright brutal at times and I would often find myself defending Henry's honor. A deep but disquieting bond would form between us.

As we walked through the jungle, I asked that age-old question, "*What's gibnut taste like*?" And no, the answer was not what you think—chicken!

At camp that night, Henry disappeared into the jungle as he so often did but this time upon his return he was carrying a freshly-killed gibnut in hand. Eager to please me, Henry had taken my question as literal.

*Wow! I really hadn't intended for him to go kill one!* I would have been perfectly happy to wait till the next time a street vendor had one cooked up. While I was somewhat troubled by the kill, I was honored by Henry's thoughtful effort to feed me some wild gibnut. He also offered me the first bites of the brain, which according to Henry, is best part of the rat. I graciously declined the offer, however.

So rat on a stick was the evening dinner. The meat is likened to pork and in my opinion, nowhere near the fine taste of lobster. But to Belizeans, it's to die for. This meat is held in such high regard that it was fed to visiting Queen Elizabeth of England, giving it the name Royal Rat. Not sure if she grew whiskers or snorted afterwards, but I was told she did make squeaky sounds when she walked. I'm sure the Emperor was unaware of what she had just eaten.

Anyway, gibnut is reasonably good, not great but the experience of sitting around a campfire and sharing a meal of wild game with the natives was an extraordinary moment in my life. We never did find any harpy eagle nests that time although later searches would end with success.

When my good friend Joel Beverly came to visit us in Belize, he insisted on trying some of this uncommon rodent. While we searched high and low for a restaurant serving of grilled rat none could be found. He would come to coin the phrase "*Give it up, give it up gibnut*"—a slogan I still use today.

The green iguana, *Iguana iguana,* also known locally as bamboo chicken is a large lizard (up to 7 feet) that spends most of its time basking in trees located along river banks. The oversized dulap of male iguanas is used to attract females or to claim territory. Adults are largely herbivorous while juveniles are omnivorous, having a more varied diet. Interestingly, this species is under threat of disappearing in many locations, due in part to habitat loss, but more importantly as a result of over-harvesting of female iguanas and eggs for human consumption!

In south Florida where this large lizard has been introduced, its numbers are completely out of control. But one way to deal with an invasive species is to eat it, which is what they do in Belize with the exotic lion fish. Perhaps south Florida should take a page from the Belize book on dealing with invasive species; *If you can' beat them, eat them!*

Bamboo chicken is occasionally available at restaurants here in Belize where I have tried it and, yes it does taste somewhat like chicken. Live iguanas are occasionally sold by kids along roads especially near Maya Villages around Easter, when females are gravid (full of eggs). Unfortunately, the sellers, in an attempt to prevent their wares from escape, cut the tendons of these creatures and use it tie their hind legs.

Feeling bad about my lizard dinner the night before, I stopped and bought four live gravid females at one of these roadside stands with the intention of releasing them unharmed back into the wild (even with cut tendons they will survive).

Feeling good about my rescue—wouldn't you know it the next day those darn little Maya girls were at it again, probably with the same ones I had bought the day before. I learned that my feel-good moment had only perpetuated their zeal to sell more lizards.

Oh well, maybe it was better to have lived one more day in the wild than to have been eaten the night before.

# DAK Attack

"It was another dark and stormy night"—no not the beginning of another love-puke novel but a nightly ritual in the rainforest. We were all in our jungle hammocks dreaming of dry weather and any meals that didn't include rice and beans, when a clattering of pots and pans abruptly woke me from a deep slumber. Something big was walking around our camp. I could see a dim glow in the distance but still couldn't quite make out the shadowy figure.

*Was it a tapir, a mountain lion or perhaps a just another curious jaguar?*

While the jungle usually sleeps during the day, at night it comes alive with slithering serpents, spiders the size of hamburgers, and cockroaches that could bite your head off—you know the kind of stuff nightmares are made of. While your skin is doubtless crawling with horror at the thought, my skin is often covered by these creepy crawlies. I'm not kidding, I live for this stuff! Another pot fell and that dim glow got brighter and brighter. *What the hell was going on out there?* Someone had started a small fire but why at that hour of the night?

It was bottomless-pit Henry, having a DAK attack at 2:30 in the morning. This guy never stopped eating, yet he was skinny as a rail. He had started a fire to cook his smuggled in can of DAK, a low-grade version of Spam, or container of nauseating cuts of pork (eyelids, ass, and lips), mass-produced and loaded with those heart-healthy chemicals like sodium erythorbate, sodium nitrate, and sodium phosphate that a dog wouldn't eat! This is the stuff that dreams are made of if you're a 30-foot long tapeworm in someone's intestinal track—Henry's perhaps?

I have to admit that the aroma of roasting DAK brought back memories as a Boy Scout—cooking that nasty Spam over a campfire. Deep in the jungle, even smoked eyelids, ass, and lips sound better than endless meals of rice and beans!

## The Grub Hub

Although DAK and Snack Attacks were common at BFREE, there are even less palatable foods and smells that can turn your mouth and nose every which way but loose. One person's culinary and sniffing delight can be another's gastronomic and nostril disgust. Like smelling your own farts, which I personally enjoy, even thou its goes against societal norms. Nevertheless, when opportunity comes knocking to wipe the permanent smile off your visiting nephew face, you also have to take one for the team.

While on an early morning stroll in the jungle, my nephew Aaron and I came across a rotting log with edible displeasures. As I dug through the decaying wood looking for salamanders and snails, I came across several large beetle larvae, also known as grubs. You know, those white plump kind that look like an over-inflated balloon ready to burst. They were filled with white and brown stuff, glistening in the sun like two Krispy-Kream cream-filled donuts. Yummy!

So what do two crazy guys from Kentucky do for jungle entertainment? Its simple. A culinary challenge of dueling palates.

"I'll eat one if you do," I challenged, not really expecting him to take it up. But to my surprise, Aaron somewhat reluctantly said, "Okay." He wasn't about to let his eccentric uncle outdo him! Being the senior crazy, I would go first, followed by the lesser, younger version of crazy of the two, Aaron. *But wait, we forgot to squeeze out the half-digested, smelly, and fermented wood fiber first.* By then, it was too late! The grubs had been chewed and swallowed.

The taste was downright awful, tasting a bit like a cockroach smells after its been smashed. The fermented rotting wood left the taste of the grub in our mouths that would last for days.

Over the years, I have eaten fried grasshoppers, fish eyes, sautéed cicadas, raw ants and their eggs, and have wolfed down a cupful of termites yet none of those were all that unpleasant tasting. But wood-filled grubs from rotting logs were the worst of the worst in terms of taste, smell, and texture.

Nevertheless, the memory of watching Aaron's face go from unending smile to descending frown was priceless and well worth the nasty taste I endured for days.

# Gold Diggers & Bush Meat

After some time at BFREE, I would be put in charge of a bird team made up of people from surrounding villages including—you guessed it DAK-filled, gibnut-hunting Henry. Our research would involve catching birds in mist nets and recording data such as species, age, and other general information. Jamie Rotenberg, a professor at the University of North Carolina at Wilmington and bird expert, would train our team. From BFREE, we would travel up into the Maya Mountains where much of the research was conducted.

Remember that deep but disquieting bond that Henry and I had? It would wear thin in the upcoming year. At times, it seemed that the only friend Henry had at BFREE was me. Without a doubt, Henry's immature and childish ways brought him much ridicule from his peers. He would often disappear without a trace and then suddenly show up out of nowhere. He was a hard person to keep tabs on—kinda like the "Where's Waldo" of Belize. Much of his disappearing act was a result of being incredibly homesick, missing his family, and when it got to be too much, he would just up and leave, telling no one. We finally had to let Henry go. Nevertheless, Henry continued to visit me at BFREE.

On one of our last days before returning home to the states to visit family, there at our doorstep at the crack of dawn stood Henry, bearing gifts of gibnut and turtle shells. It was indeed a singular moment and quite honestly, one of the nicest things anyone had ever done for me. It was no small task since Henry had to begin walking at 3:00 AM to make the two-hour trek through a dark jungle from his village to reach our house. But this would not be the last time I saw Henry.

Six months later, he showed up out of nowhere asking for his job back. And while I wanted to help Henry, the team wouldn't have it. So, disappointed, he offered to sell me a scarlet macaw for 3000 Belize dollars, which is around 1500 US.

As you might imagine, I was stunned and horrified since this scarlet macaw species is a highly endangered bird in Belize with around 200 remaining in the wild. Remember, Henry was part of the bird banding team here at BFREE and should have known better, or so I thought.

*"How did you get this bird?"* I asked.

He responded without remorse, "A rock from a slingshot to the head which knocked it unconscious, but it's ok now."

While I tried to wrap my head around Henry's motives, it was, in fact, as pure and simple as Henry's mind. Selling a scarlet macaw would feed his family for a month. There was no scolding this adolescent for the odious crime and there would be no repercussions. The next day, I contacted the authorities who came to pick up the bird, and Henry would go unpunished, unable to pay any fine implied.

This is common practice in Central and South America where people are just barely squeaking by. So, they sell artifacts dug from Maya ruins, turn to the forest for bush meat (sources include everything from squirrels to monkey's), cut valuable trees like rosewood to build our fancy tables, tear open the land for gold to make our wedding bands and build up our stockpile of gold coins, catch and sell wildlife so we can brag about our exotic pets, clear pristine rain forest to raise cows and soybeans for our hamburgers and soy dogs, or replace century-old jungles with endless monocultures of palm oil trees so that peanut butter doesn't stick to the roof of our mouths.

And so goes the slow grind of planet Earth and its finite resources, all fueled by civilization's hunger for shit they don't really need. Wealthier countries are the biggest importers of wildlife, which in general, originates from poorer countries. The USA, France, and Italy are the chief buyers, while Indonesia, Jamaica, and Honduras are the leading wildlife exporters. More than 4 million wild-caught individuals from 12 animal groups were legally traded across international borders between 1998 and 2018.

More concerning, though, environmental protection agencies and wildlife charities estimate that up to 350 million illegal specimens are bought and sold on the black market each year, estimated to be worth $19 billion annually. But an ever-increasing human population has to feed itself somehow.

Although we are quick to take up political arms and trash-talk South Americans for cutting and burning down the Amazon rainforest and committing genocide on indigenous tribes, we forget that our US history is fraught with early white colonizers who also cut down endless swaths of virgin forest, while killing and displacing earlier Native Americans. Ironically, if it came down to a choice between providing basic needs for ourselves and family, we would all take these same actions.

In the US, we have the comfort of prosperity, so it's no skin off our teeth to protect parks and wildlife. Most Americans live privileged lives, not giving much thought to the cost of our lifestyles on the environment, especially if it's worlds away.

While the disembowelment of Earth's limited resources is largely carried on the backs of the poor, it is primarily driven by the gluttony of the privileged. Slowing down or stopping demand for these products will help, but people still need a means to live. One thing is for certain, business-as-usual will only end badly for us all.

## Where There is No Doctor

"Where There Is No Doctor, A Village Health Care Handbook" is an unimaginable book by David Werner. It's a must-have text if you live days or even hours away from a medical facility or doctor. The information and illustrations in the book are priceless and can help cure pain, suffering, and even save lives.

At BFREE, we would often turn to this unconventional text to solve tropical health mysteries and emergencies. The illustrations are so bad they are good, even admirable in their own convoluted way! Clearly, the artist paid more attention to the medical affliction and injury drawings than the people drawings that look more like aliens or the walking dead. The illustrations are definitely my favorite part of the book and worth buying just for that alone.

While at BFREE, I would hold two titles: Resident Biologist and Camp Doctor. In a past life, I was trained as an EMT/paramedic during my time living in Cincinnati, Ohio where I worked as a part-time paramedic on a life squad and full time Emergency Room (ER) Technician at Providence Hospital. My duties in the ER would include taking vital signs, starting IVs, suturing, intubation, defibrillation, CPR, assisting in bone-setting, and minor surgeries. So when I wasn't chasing snakes and crocodiles in the jungle, I was playing doctor at BFREE. In Belize, until recently, you could buy just about any drug you needed including Lidocaine at the local pharmacy—which I kept in stock along with sutures and tools needed for minor torn flesh repairs.

My doctor skills were tested our first year when late one night Steven Brewer, an outstanding botanist and long-time fixture at BFREE, came staggering in from a solo week-long expedition in the jungle with a bloody hand wrapped in rags. He had slipped and fallen on his razor sharp machete, receiving a deep laceration between the right thumb and index finger. Machetes are the tool of choice and a must for any journey into the jungle but account for the majority of lacerations in Belize.

Steven is a rugged outdoorsman and tough as nails but when it came to his laceration and blood, he was like a newborn puppy. I don't blame him, the cut was deep and nearly to the bone. And while I can paint the side of a barn with someone else's blood, I become a total cupcake seeing my own. But he needed stitches, so I proceeded to infuse Lidocaine into the gaping hole, but not without Steven first chugging a bit—ok, a LOT—of Belizean One Barrel rum.

As the Lidocaine and liquid courage began to take hold, Steven started to relax. I then dove into the gaping wound and began to look for tendon damage, which to my surprise, there was none. After all, any tendon damage would need to be repaired by an expert at least 8-10 hours away. Steven was indeed lucky that night. There was, however, severed blood vessel damage and swelling that would retard the healing process.

With the wound stitched up, the swelling got increasingly more severe. I needed to siphon off the surplus blood. Where there is no doctor, leeches would do the trick, so off to the swamp I headed to search for the finger-saving bloodsuckers.

These aquatic vampires could be applied to the swelling thumb where they would attach themselves, removing the surplus blood. When veins are cut off, the incoming blood cannot return to the heart, thus causing swelling and additional tissue damage.

As it turned out, the swelling subsided without the need for leeches. The somewhat disappointed Steven ended up with around a dozen stitches but no worse for the ordeal except perhaps a greater respect for that impartial flesh-slicing machete.

Another suture-fest would include the time Katelyn cut her finger with—you guessed it—another sharp machete. With no Lidocaine left to numb the site, we turned to ice which has a similar deadening effect.

My skills as a jungle surgeon would be refused when one of our visiting Mayan workers split his nose wide open by a falling branch while working in the cacao plantation. The staff brought him to me for treatment but when I pulled out the needle and thread, he wanted nothing to do with this witch doctor from Cincinnati, Ohio. He quickly retreated back into the jungle never to be seen again, later sporting what must have been an impressive scar across his nose.

Living in such a remote place where there are so many things that could go wrong at any given time, it is amazing that we didn't have more medical emergencies. But if we had, we were well-equipped, *living where there is no doctor.*

# The 12 Death Gods of XIBALBA

When Judy and I first came to Belize, we would be bewildered by the amount of intact ceramic vessels found in caves throughout the country. Any cave you could stand up and walk through would shelter dozens of ornate clay pots, arrowheads, even human remains. How was this possible? Back home in Kentucky, nearly every rockshelter that was ever occupied by early Americans had been raped and pillaged of its artifacts.

But in Belize, there are hundreds of caves with thousands of pots and other artifacts exactly where the ancient Maya left them more than 1,000 years ago. We couldn't believe that these relics were not removed and sold on the black market?

The dreadfully dark and even sinister answer could be summed up in one word, **Xibalba,** the place where the 12 death gods of the Maya purportedly resided in dark underworlds. Even to this day most Maya fear this place of the undead and remain afraid to enter caves. This fear, however, is slowly eroding away being replaced instead with a desire to gather and sell these artifacts for basic needs.

Belize is blessed with endless cave systems, including one of the largest on the planet, Chiquibul Cave, with a room the size of a football field. With all these deep dark recesses scattered throughout the country, it's not surprising that the grip of superstitions surrounding Xibalba ruled those early Maya.

To understand more clearly just why this place, Xibalba, wrought such fear in the minds of the Maya, one needs only to learn just a bit more about the 12 death gods who supposedly dwelled in these mysterious places. The dark, gloomy, and ultimately terrifying underworld called Xibalba (she-ball-buh), roughly translated means "place of fear." It was believed to be ruled by twelve bizarre-looking death gods and, according to Maya legend, located in a wet courtyard below the surface of the Earth, where torture and death were entertainment of these freaks. The first among the Maya Death Gods ruling Xibalba were Hun-Came ("One Death") and Vucub-Came ("Seven Death"), though the former was the senior crazy of the two.

**The Lords of XIBALBA**
*Artwork by Diego Rivera*

The remaining 10 gods were often referred to as demons and given domain over various forms of human torture and suffering such as sickness, starvation, fear, destitution, pain, and ultimately death. Sounds like the God of the modern day Christian Bible. These sick bastards worked in pairs:

**Xiquiripat** (Flying Scab) and **Cuchumaquic** (Blood Gatherer) who sickened people's blood.

**Ahalpuh** (Pus Demon) and **Ahalgana** (Jaundice Demon) who caused bodies to swell.

**Chamiabac** (Bone Staff) and **Chamiaholom** (Skull Staff), who turned dead bodies into skeletons.

**Ahalmez** (Sweepings Demon) and **Ahaltocob** (Stabbing Demon), who hid in the unswept areas of people's houses and stabbed them to death. Perhaps this is why Mayan woman are up at the crack of dawn, sweeping dirt floors you could eat from!

**Xic** (Wing) and **Patan** (Packstrap) who caused people to die, coughing up blood while out walking on a trail.

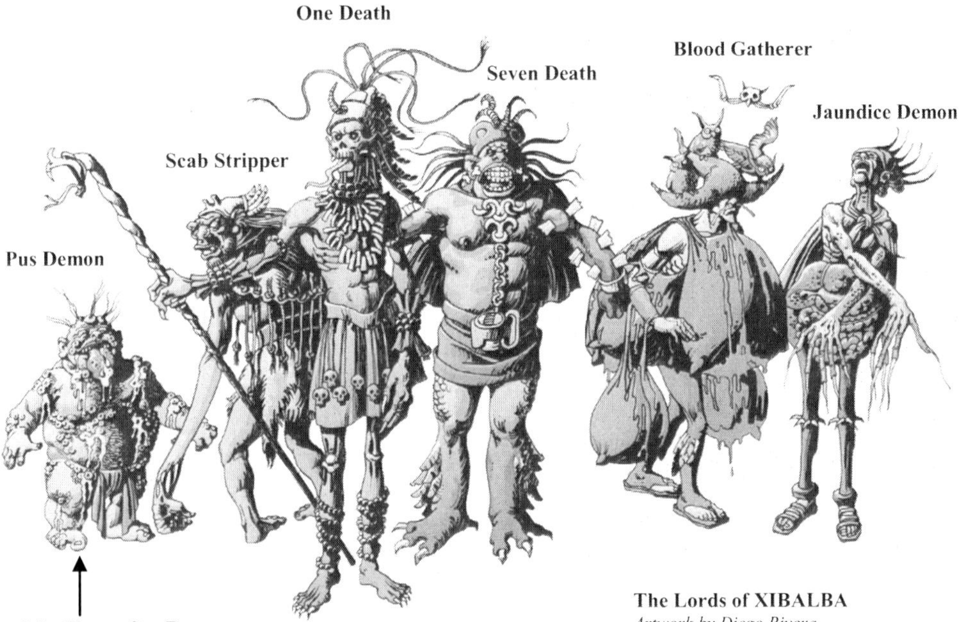


**The Lords of XIBALBA**
*Artwork by Diego Rivera*

The remaining lesser demonic inhabitants of Xibalba are thought to have fallen under the dominion of one of these Lords, going around to carry out their sick and twisted duties.

Xibalba was a great underground town full of trials and traps for any poor schlub that wandered into the city which included at least six deadly houses. The first was **Dark House**, a house that was completely dark inside. The second was a **Cold House**, full of bone-chilling cold and rattling hail. The third was **Jaguar House**, filled with, you guessed it, hungry jaguars. The fourth was **Bat House**, filled with dangerous shrieking bats, and the fifth was **Razor House**, filled with blades and razors that moved about of their own accord and finally a **Hot House**, filled with fires and heat (Hell?). Sounds like great places for a Halloween Party, doesn't it?

One of our most memorable research trips into the jungle to study snails took us past and through the underground worlds of Tiger Sandy Bay in Central Belize; a place honeycombed with hundreds of caves. There were dozens, if not hundreds, of vessels of all shapes and sizes, untouched for centuries but nowhere to be found were the ruthless demon gods. We imagined a time when hunters returned with gibnut and brocket deer while gatherers cooked palm hearts in large pots over leisurely fires. It would have been a time of plenty until it wasn't. The collapse of the Mayan Civilization was inevitable with an ever-increasing population tied to the dwindling resources. But their great stone cities would survive the test of time, as did their ceramic vessels in hundreds of scattered caves across Belize.

Juan Cho, who is Kek'chi Maya and a dear friend, had discovered a 1000-year-old Mayan altar in a cave deep within the Maya Mountains and agreed to take us to the site. Several weeks earlier, Juan had traveled to this sacred site with some Mayan friends from the local village, which he would later regret. He described it as a jaw-dropping place of wonder, the most amazing intact Maya History he had ever seen. The walls were decorated with jaguar skulls, jade artifacts, painted pottery, and, what appeared to be a ceremonial altar.

We were anxious to experience this extraordinary place as well, so for the next three hours we hiked through a trail-less jungle to the cave. When we finally reached the entrance, Juan noticed something out of place. As we entered into the cave, the desecration of the site would become increasingly obvious. We were now standing, not in a great room of Maya history, but a place that had been ransacked just days before. The exuberance of the site had faded. There were broken pots and effigies scattered everywhere. The sacred site had been ravaged for mere pennies.

We watched our friend Juan fall to his knees in tears, as he scanned the vandalism that lay before us. It reminded me of a line from an old John Denver song;

"*There are those who would deal in the darkness of life, there are those who would tear down the sun, and most men are ruthless but some will still weep when the gifts we were given are gone.*"

Embezzled was 10,000 years of history, wiped clean in a moment of uncertainty. Why superstitious Maya would risk the wrath of **Xibalba** and the 12 death gods remains unknown, but perhaps it was quite simple.

While these stolen artifacts could easily be chalked up to just basic greed, there's another side to consider. This stuff technically belongs to the Maya as part of their heritage and, therefore is theirs to do with what they want. And if they decide to sell the artifacts for money to buy food to put in the bellies of their children, or bolster their basic lifestyles, who are we to say this is wrong action?

After all, this is what European and American cultures have done ever since we decided that owning land and stuff gave us power over people. So our shit has been passed down or sold for generations, keeping us all wealthy, in terms of money that is!

I have talked with some of my Mayan friends who are frustrated that much of their cultural heritage was "legally looted" by foreign archeologists who dug it up, stuck it in boxes and sent it afar, where it now sits in universities collecting dust or in museums worlds away, for people to gawk at. But if the Maya dig artifacts and sell their rightful inheritance, they are severely scolded, fined, or even jailed.

I used to think that looting archaeological sites was black and white, never questioning an archeologist's intentions when it came to digging for history. If you dug a site for science it was ok, but if you dug a site for monetary gains, it was wrong. It was that simple or so I thought. But nothing is ever that simple.

No doubt archeologists reading this are probably freaking out about now. But as one of my close archeologist friends would say to me when I tried to explain to him the importance of salamanders in ecosystems, "*So what?*". He was probably right on a number of echelons.

No one would disagree that digging archaeological sites hasn't contributed a wealth of information to human history. But it seems that no matter what we discover about our past (fallen empires, wars, famine, disease) or what we know about our future (climate change), we all just turn a blind eye and continue grinding what's left of planet Earth to dust.

Has information about our past really moved humanity to a better place? It would appear not! If past artifacts can help better the lives of the Maya today, why not? If archaeology tells us history is bound to repeat itself, nothing new here! If you're an archaeologist coming to Belize to dig up the past, make sure you first ask the rightful owners before you steal their inheritance.

Obviously, I care more about the living and its future than the dead and its past. For me, it all boils down to being practical in a world gone mad. At the end of the day, it makes more sense to feed the hungry than to feed our chronological egos.

And for those of you that think I care only about protecting my field of study (biology), check out my ill-fated, biological reality, under "**Current Environmental Views, the Inconvenient Truths of Tomorrow**". For a secular, non-Dan point of view, go to "**Mongabay**" for a dose of reality with regards to our crumbling planet or watch "**Don't Look Up**", a (2021) movie by Leonardo DiCaprio. Both illustrate very well our current and dishonorable veracity!

# Long Beautiful Hair

♪*Give me hair, long beautiful hair, shining, gleaming, streaming, flaxen, waxen down to there, shoulder length or longer.*♪ a great line from an old long-hair hippie song from the 1970s when it was cool to be a hippie, a song that played in my own personal headspace soundtrack from the moment I met an extraordinary Belizean.

Naturalized Belizean Gomier Longfellow, a dear friend and Master Vegan Chief extraordinaire who lives in the seaside town of Punta Gorda, an hour south of the BFREE entrance road, possesses mammalian locks that should be in the Guiness Book of World Records.

There's long hair, and then there's Gomier's magnificent hair. From the day he was born, according to the man himself, he had never cut his hair and it's been growing ever since.

It's long, really long, longer than long, the longest hair I have ever seen on a man or woman combined. His massive dreadlocks are generally covered by a rastacap that could hide a 20-pound gibnut. In fact, our numerous excursions to consume his incredible culinary delights notwithstanding, we had never seen even a glimpse of the massive locks hidden under that cap, that is until one day when I pressed him to reveal the secrets hidden beneath that red, black, yellow, and white woven cap.

*"Are you sure you are ready for this?" he said grinning broadly.*

And before I could nod yes, the cap was flung off and the fury of dreadlocks was released. Gomier let it all hang out and hang out it did. My jaw dropped in disbelief at the yards, not inches of gold and black hair falling to the floor and beyond. You heard me right. This 6-foot-tall man's hair hung down to the floor and continued on like a Persian rug for another 5 or 6 feet behind him. It was truly impressive!

His hair weighed more than I did and although Gomier was a thin man of stature, his neck muscles must have had the strength of a Ceiba tree to support such a weight. To this day when I hear that song, "Long Beautiful Hair", I think of Gomier's unending mass of hair.

While his hair is the thing for the record books, that is by no means the only redeeming quality of this soft-spoken man who has devoted his life to promoting the healthy consumption of quality foods. His tiny roadside restaurant wasn't much to look at but the food is what us foodies only dare to dream of. His commitment to the promotion of whole grains, natural probiotics, non-dairy, vegetarian, or pescatarian food consumption is something of an anomaly in a country where white flour and white rice rule the ever-increasing waistlines of Belizeans.

His simple food philosophy "**Health is Wealth**" has been promoted throughout southern Belize as he traveled to schools to preach the word of healthier eating habits with the zeal of an evangelical preacher!

Truer words have never been spoken! For without our health, we are indeed poor in body and mind.

Meeting Gomier, a man who shared our commitment to consuming the healthiest food on the planet, in the tiny town of Punta Gorda added greatly to our lives, *living where the cats eat the dogs.*

# Caracol, City of Snails

Caracol which means "snail" in Spanish is the name given to one of the largest ancient Mayan archaeological sites in the world, located in what is now the Cayo District of Belize. The ancient city still stands today, although largely covered in jungle. It sits around 40 kilometers south of the bustling town of San Ignacio, resting on the Vaca Plateau in the foothills of the Maya Mountains. Caracol covered approximately 200 square kilometers of land, an area much larger than present-day Belize City and supported an even larger population of between 120,000 and 180,000 people at its peak. Caracol has approximately 267 limestone-structures per square kilometer, 85% higher than the great rival Mayan city of Tikal in Guatemala.

The road to Caracol is a long and arduous journey, not soon forgotten. On a dry day, it takes around 3 hours to reach the site and on a wet day, meaning torrential rains, nobody in their right mind travels it. It's 35 miles of gravel the size of oranges and bowling balls which will turn your innards every which way but loose. There's no shortage of mud and rickety bridges to cross either. If you ever wondered what a milk-shake feels like, drive the road to Caracol.

While BFREE has the "road of lost soles (souls)"; Caracol has the "road of pounded inners". It's also the road having the best chance of being robbed by Guatemalan bandits. We were always told not to stop for anyone except the Belizean Defense Force and the few times we traveled it, we heeded their advice. As with many things, sometimes the risks are worth the rewards, especially to a zany snail biologist conducting research for that spellbinding and illusive creature. I had no choice but to visit the greatest Mayan ruin of all time, its name honoring the greatest animal to ever walk—well, actually crawl—the Earth.

Armed with as much information to keep us as safe as possible, Judy and I headed out to Caracol. It was a misty morning in the mountains of the Maya as we bounced and bobbed up and down, sideways, and sometimes no ways; a tedious crawl over a road paved with butt-deadening stones from hell, perfectly designed and placed on the road to give maximum discomfort. Few live beyond the town of San Ignacio for good reason. It is without question the road less traveled, usually my kind of road, but even I would come to curse this teeth-rattling, brain-jarring journey.

After nearly 3 and a half hours of what seemed like a road to nowhere, we finally arrived. As we entered the park we caught glimpse of a large tannish figure crossing the road, and in just two colossal leaps, disappearing into the jungle. WOW, a mountain lion, as we sat in disbelief, was welcoming us to the lost world of the Maya.

It felt surreal to be in a place named for the mollusks I had devoted much of my life studying. Would it live up to its name? As we explored the many ruins, I thought how ironic that this great city was built from the shells of snails and other calcareous organisms.

Let me explain. In the past, the colossal accumulation of deceased mollusks (including marine snails), corals, and tiny creatures known as Foraminifera (which have calcareous skeletons), provided the necessary building material to create limestone. The same limestone used to make cement and form the foundations of our cities and homes also supplied the necessary building material used by the Mayan to erect their immense shrines.

We had a terrific time—finding many new species of land snails scattered throughout the ancient ruins. On the return trip, not only did the road batter our bodies again, it destroyed our truck as well. About half way back from Caracol, the front right wheel of the truck just fell off.

There we would wait for hours for a tow truck to carry us back to San Ignacio. In Belize, slow and laissez-faire is the name of the game, not the hurry and worry that's so common in the US! Belize is no place for the inpatient. Our down time allowed us to reflect on the extraordinary City of Snails.

As I contemplated our visit to Caracol, these words came to mind.

*"In the shadows of great temples I stand in awe, not to the stone works of man built for torture and gods, but for momentous works of time and evolution I hold in my hand. The study of snails is an inquiry into subtlety and masterful sculpture".*

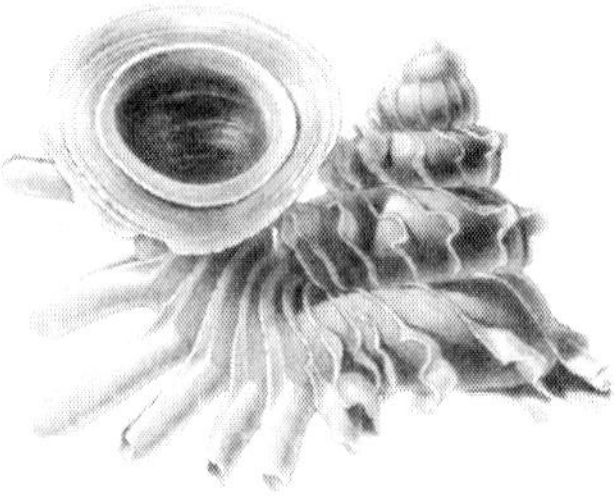

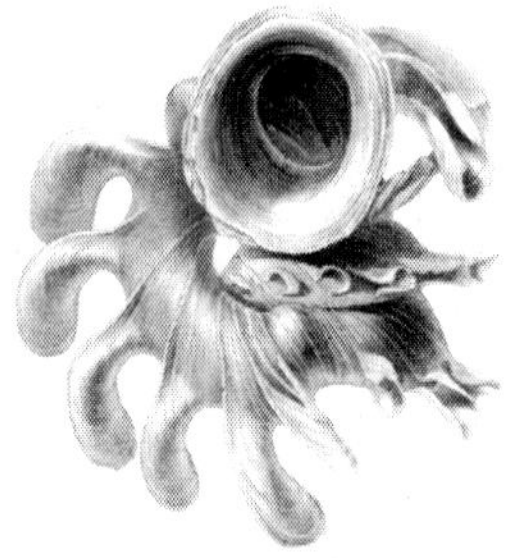

Illustrations by Jaap Verneulen

# Little Amazon

There's a small river in southern Belize that runs the entire border with Guatemala called the Sarstoon. To the north, the River is bordered by the Sarstoon Temash National Park in Belize and, in Guatemala, to the south lays mostly broken jungle and small family farms along its banks.

But as one travels upriver, tiny homesteads give way to an endless sea of jungle. Few clear-skins, as Caucasians are called in Belize, have ever ventured into this unexplored wilderness. The trip was reminiscent of another river trip we took years before up the Amazon in Peru to the Tahuayo River, minus pink dolphins and anacondas.

Our goal? To survey this remote area for land snails, what else? Our research team including Ron Caldwell, (snail biologist & snack master), my adventurous wife, Judy; Belizean birder and all-around extraordinary tour guide (who thankfully spoke Spanish), Wilfred Mutrie. We would leave Punta Gorda at sunrise aboard a 20-ft fiberglass skiff with a canvas bimini top for the trip south in the Bay of Honduras along the coast to the mouth of the river.

As we scooted along, up over the horizon, the faint image of another boat appeared. It was headed toward us as fast as we were headed up river toward it. Wouldn't you know it? The first humans we met on the river were not just local fisherman but Guatemalan soldiers wielding AK-47s and attitude. As we got closer, Wilfred, our skilled boat captain, said.

"*Be calm, that boat is the Guatemalan Coast Guard. Don't say anything; Let me do all the talking.*"

Well THAT would be easy, since the only other person on the trip who spoke Spanish was Ron but his Spanish-speaking skills were more like Spanglish!

With guns locked and loaded and pointed in our direction, they asked in Spanish, "*What's your business here*?" (probably thinking we were drug runners, a common activity along the southern border of Belize.)

Wilfred responded, "*We are researchers going up the Belizean side of the Sarstoon to look for land snails*."

With confused and doubtful looks, scratching their heads and a brief pause, they answered, "*Nails, there's no nails up there, only jungle*."

Right about now I was wishing I studied something like jaguars or tapirs instead of snails! Why? Because explaining to someone that you study land snails is like saying, "*I'm my own grandpa*". It takes forever to sink in. Once it does, the next question is invariably, "*Why would anyone in their right mind study snails*?" It's hard enough to explain this unusual research subject to scientists. Now imagine trying to explain snail research to four overly-paranoid, gun-slinging military types. It would have been easier to convince them that we were kangaroo farmers looking for that one that hopped away.

Still confused, with itchy trigger fingers, I was sure they were probably even more convinced that we were drug dealers after what probably sounded like a load of horseshit. With skeptical looks on their faces, the lieutenant, clearly the one in charge, gruffly asked for our IDs and other paperwork. Fortunately for us, Ron had dotted all the "i's" and crossed all the “t's” when it came to our research trip.

"Here’s our permits sir, and see, it states we are conducting land snail research in Belize. Besides only crazy people would make up such a silly story and we’re not crazy just unconventional," I explained and Wilfred translated. Since neither Judy nor I knew Spanish, we are really not sure what Wilfred ACTUALLY told them but whatever he said, it worked.

We would eventually convince them that we were in fact legitimate. But you could see the disappointment on their faces. *Damn, no drug bust here or gringos to haul in and soak for some extra cash.* So they finally let us all go. Still a little unnerved by this brief encounter with Guatemalan military in an area that is still contentious between Guatemala and Belize, I thought, "And most people think there's nothing dangerous about studying land snails!"

Up the river we travelled while making random stops along the way to collect that most riveting group of organisms. But there would be one more stop before returning home. An unexplored cenote I found while searching Google Earth maps.

Our guide, Wilfred said the locals avoided that place at all costs. Legend had it that the cobalt blue waters were bottomless and a place of lost souls that can be seen flying around the water's edge during certain times of the year.

While the idea of lost souls was intriguing, I surmised that there was in fact a scientifically-based explanation to this phenomena—the helicopter damselfly—the largest dragonfly in the Americas with a wingspan of more than 6 inches. An inhabitant of mature forests, eggs of this large damselfly are deposited in bromeliads and tree crotches that hold water. Nymphs feed on mosquito larvae and tadpoles of frogs that lay eggs in these container habitats. As it hovers, the adult damselfly feeds on kleptoparasitic spiders snatched from the webs of orb weavers. Interestingly, the Maya believe that these large damselflies are the spirits of recently-deceased humans, insisting that they not be caught or harmed.

After a short walk from the river, we did find the cenote minus any floating spirits or damselflies, but impressive nonetheless. Similar to the famous Blue Hole of Belize located in the Caribbean Sea which is 410 feet deep and 948 feet wide, this cenote is about half as big, but its depth is still unknown. Nevertheless, the mystery of the Sarstoon Cenote lives on and remains unexplored to this day. Who knows what truly lays at the bottom of this great abyss? Perhaps it is the entrance to the lost underworld of **XIBALBA**? The trip up the Sarstoon became just one of the many memorable research expeditions that would be forever etched in our memories.

# Oak Ridge Expedition

Most of my research expeditions into the Bladen Nature Reserve were focused on the lowland tropical rainforest. But the expedition to the high-elevation forests of Oak Ridge presented an entirely different habitat with new challenges and discoveries to be made.

The journey would take us to an unexplored and unstudied region deep in the Maya Mountains. But getting there was no easy task. It would take our research team almost three trailless days to reach this mysterious place. Oak Ridge sits at the top of the Maya Mountains in some of the most intact rainforest left in the Americas. At an elevation of around 3200 feet, it just barely qualifies as Montane or Cloud Forest. Oak Ridge, named by Steven Brewer for the massive tropical oak trees and golf ball-sized acorns, was a totally different world compared to the tropical rainforest below. The impetus for this treacherous journey was a multi-organism study with our team made up of bat biologists Mark Gumbert and Price Sewell, botanist Steven Brewer, and a gang of Mayan porters.

While the rains were relentless (which I was somewhat accustomed to by then, having lived in the jungle for several years), I was not prepared for some of my coldest camping nights ever. Jungle hammocks are great at keeping you dry but lousy at keeping you warm. Nights are long when you're cold and days seem endless from bounding rains, giving us all a ghastly case of "wet bone".

But the miserable nights shivering in my jungle hammock were richly rewarded by the discovery of several new species of land snails and the wails of cacomistles. This uncommon mammal is absent from jungles below, so most Belizeans have never heard—let alone seen—this amazing animal. The calls of this arboreal creature are unlike any others in Belize and are heard as the species runs through the treetops with great agility. Even where they live they're usually only seen as shadows in the mist.

Two land snails completely new to science were documented on this bone-chilling excursion; one I later named after American actor, Leonardo Dicaprio, (**Oak Ridge Teardrop, *Cecilioides dicaprio***), honoring his work in film bringing attention to the challenges facing our natural world and his interest in ecotourism in Belize. The other new species of land snail I later named was (**Montane Splinter, *Rectaxis breweri***), in honor of Steven Brewer, an extraordinary and passionate botanist who has spent countless hours exploring and cataloguing Belize's outstanding plant life.

Emphasizing the unexplored vastness of the roadless and trailless Maya Mountains, other important discoveries documented at Oak Ridge included two new species of herpetofauna, a tropical kingsnake and salamander (both which remain undescribed), new plants for the region, and previously unknown feeding behaviors for the fringed-lip bat.

In spite of all these amazing discoveries, after two weeks of wet bone, rice and beans, and cold, sleepless nights, our weary team of zany biologists were ready to come down out of the clouds to the steamy sweat -box below.

Yet, with all the discomforts of the expedition, I consider myself very fortunate to be one of a handful of modern humans to experience this unique, and at times, intimidating habitat.

# Blue Creek Cave

In extreme southern Belize, only 30 air-miles from the Guatemalan border, sits the Mayan village of Blue Creek. While the tiny town looks like any other Mayan settlement, its backdrop is nothing less than spectacular; a chain of steep-sided hills, sinkholes, outcrops of limestone, and a mammoth opening called Blue Creek Cave.

Whereas this cave is an impressive sight to behold, just locating the path to reach it can be challenging. With only faded hand-painted signs to mark the path, it's easy to miss it entirely if you don't know exactly what you are searching for. Upon parking, it is almost guaranteed that before one can even get a daypack on, you will be surrounded by beaming Mayan girls no more than 9 or 10 years old, armed with backpacks full of trinkets, jewelry, and other pieces of handmade art, and an outstretched hand ready to relieve you of your US dollars.

After wading through the bevy of budding entrepreneurs, a rocky path winds its way along what seems to be little more than a meandering stream. A little less than a half mile away, a visitor will reach the home of International Zoological Expeditions (IZE), a privately-owned and managed research and environmental education facility with cabins and a central kitchen, similar to BFREE, all situated beside Blue Creek.

A short distance past IZE facilities, the vast darkness of the enormous cave bursts into view. The entrance to Blue Creek Cave is so large you could drive a tractor-trailer into its opening. The large creek begins worlds away on the other side of the mountains in an isolated valley. But, instead of finding an easier path by flowing around the hills like most respectable waterways, over eons of time, Blue Creek carved a mile-long tunnel right through the mountain's center.

What remains is an immense cave system with a vast network of pits, waterfalls, and enormous rooms filled with 1000-year-old Mayan pottery encased in stone from the endless trickle of calcium-carbonate that hardened over time. Left are only vague shapes of ceramic vessels. The karst nature of the area was intriguing for a land snail biologist as these organism's are abundant in areas with generous quantities of limestone that provide the necessary calcium carbonate for shell building.

During our seven years in Belize, I would make multiple trips to this spectacular place. My expeditions both inside the cave's deep passages and in the surrounding jungle would lead to the discovery five land snail species unknown to science—two of which I described formally, and three yet to be named.

The (**Mayan Cave Snail, *Leptopeas corwinii***), was named in recognition of Jeff Corwin, a conservation biologist who conducted research at Blue Creek Cave for a Master's degree and continues to educate the public about the natural world through his outstanding wildlife films.

The (**Blue Creek Cave Snail, *Leptinaria doddi***), I named in honor of Frederick Dodd, founder of International Zoological Expeditions (IZE) who had the foresight to purchase and protect the wild jungles surrounding Blue Creek Cave, the only known location for the globally rare Blue Creek Cave Snail.

Blue Creek Cave is a place I hold dear to my heart for the small discoveries overlooked by many before me. Even the 12 gods **Xibalba** probably paid little attention to their presence, being more concerned with the torture of men than the affairs of snails.

# The Lost Valley

This would be the shortest of my expeditions into the Bladen Nature Reserve, lasting only three days and two nights but by far the most challenging. The Valley was a completely unexplored region of the Maya Mountains and, as such, had a high likelihood for new discoveries. The three-day trip would help determine a safe route, if any, to conduct future research.

This expedition team consisted of Steven Brewer, BFREE staff member Marcelino Pop, and me. The trek would be simple, or so we thought. From BFREE, the plan was to travel up the Bladen River, take a right turn at Richardson Creek and follow it to the Lost Valley. The first leg of the trip was fairly easy until we got to the back of Richardson Creek, where its starts to climb and climb we did. The cobble-filled creek gave way to a boulder-filled creek which then turned into short steep cliffs along the creek.

Due to the high humidity and constant rains, every surface was slippery wet, especially the butt-busting rocks. With 40-pound packs strapped on our backs, we climbed and crawled over, under, and sometimes up short cliffs, at times, reverting to a swim through deep pools with packs overhead. As grueling as this sounds and was, that would be the easy part of the trip. After what seemed like an eternity, we finally reached what would be a dead-end—a spectacular 500-foot vertical cascade of water. We could go no farther.

Undeterred by the massive wall of water, we went left to find a route around the falls. After all, we were determined to push forward and find a passage to the Lost Valley. Most of the left way was also nearly vertical but with one exception, no waterfall. So, we began the slow and strenuous climb. Fortunately for us, the cliffs were draped with lianas (vines) which became our lifelines up the steep and arduous terrain.

When we finally reached the top, the site we encountered was like something out of Jurassic Park (minus the dinosaurs). We were now at nearly 3000 feet, atop the 500-foot waterfall, looking back at what we had just hiked, crawled, and swam for miles. At this elevation, the heat and bugs of the valley below were replaced with cool, refreshing days and nights. We set up camp on top the waterfall with spectacular views of endless green valleys below. Although merely speculation, the Lost Valley, shaped like an inverted cone, may have been the remnants of an ancient volcano.

While the surrounding hills were firm to step on, the valley floor was like walking on a bowl of soggy cornflakes. In some places, the detritus (decaying vegetation) was nearly a meter deep. Although caecilians (a snake-like amphibian) have not yet been found in Belize, this is where they would occur; the soggy muck of the valley floor. Our trip would be brief and the trip back to BFREE would naturally be faster and much safer on drier rocks.

This Lost Valley would not divulge its secrets on our short exploratory trip, but its intrigue and mysteries still haunt us to this day. The plan was to return with a team of scientists to conduct a biological assessment of the entire Valley, an expedition of considerable cost and effort due to the remote nature of this hidden treasure, but this has yet to happen due to a lack of funding. So, for now the Lost Valley will hold on to its enigma.

# PhD: Piled Higher and Deeper

While I can't help but respect an individual that endures 8 years of mind-numbing classroom education, it doesn't come without creating some rather negative undertones later in life, such as a superiority complex in some who acquire this piece of paper. Whereas most PhD recipients keep this troublesome trait buried where it belongs—others use it quite liberally—wreaking havoc along the way.

I was 28 years of age when I returned to that incarceration of education—a place I hated as a kid largely due to my undiagnosed and misunderstood dyslexia. Nevertheless, I managed to survive two years in college, earning an Associate's degree in wildlife biology.

Although this was quite the accomplishment for someone who had struggled so much in a formal classroom, this piece of paper would further challenge me in unexpected and peculiar ways. My experiential learning after college would quickly surpass my college degree by leaps and bounds—placing me, rather unexpectedly, in the league of persons with much higher degrees of formal education—the PhDs. My advanced skill set would become a mixed bag of both satisfaction and anxiety.

Later in my career, I was honored to be invited to become a Research Associate of the prestigious Carnegie Museum of Natural History in Pittsburgh, a premier institution known for its extensive land snail collection and research. The request was initially based on my land snail research, published books, and journal papers on the subject. However, the initial satisfaction of such a request turned to anxiety and frustration when my CV revealed that my formal education concluded, not with a PhD, but a lowly Associate's Degree—disqualifying me for the position according to their narrow-minded bylaws. Even though my books on the subject alone would more than qualify me for a PhD level accomplishment, it was the diploma—not the experience—they required. The person who recommended me, to his credit, was dismayed and apologetic for the archaic rule that knocked me out of the position. Ironically, this same museum continues to seek my expertise on land snails.

Nevertheless, the Carnegie scrolls of elite researchers lives on, absent my name of course. I wrote back to the committee with this to say, *"It's unfortunate we live in a world where the knowledge on paper means more than the knowledge of experience"*. Albert Einstein would have been appalled at such a notion! But the scholarly snobbishness would not stop with Carnegie Museum.

Many people mistake education for intelligence, which is clearly a misstep. My Uncle Jake was one of the smartest people I ever knew, a man with a sixth grade education, but a lifetime of reading and experiences from being a farmer to an electrician, to running a theatre in a small river town in southern Ohio. He was a renaissance man in the truest sense of the word. While I agree that formal education is essential in many disciplines, it's certainly not the end all. It's but only one of many in the evolution of learning. Sadly, it is too often considered more important than experiential education.

At one time I actually considered going back to college to acquire a PhD. But it boiled down two choices: go back to school to repair and feed my damaged ego, spending a shitload of money to do so, or take the information I had acquired from years of experiential learning and put it in print. The latter won out since I had a hard time scraping up enough ego or money to get me through an additional six years of drudgery.

The most disturbing insult of my adult life would occur at BFREE while in conversation with one of those PhD professors from "Haughtiness University". As we sat talking one evening, she seemed truly captivated by my concise answers to her many questions on wildlife in Belize and other various subjects.

After an hour or so she asked me, "Where did you get your PhD?" While this would no doubt be a complement to most, I found it rather offensive. Why? Because it assumed that only someone who has spent eight years in college could possibly answer her questions the way I so correctly did. I proudly responded by saying, "I have an Associate's Degree from Hocking College, and more importantly, a unstoppable passion for learning." And its all I ever needed!!!

Seemingly befuddled by my answer she said—and I kid you not—**"And people take you seriously?"**

*Well lady, you did for at least an hour, so you tell me?* I thought. *So the only reason you think people take you seriously is because of that PHD hanging on the wall above your throne of insecurities*? I wanted to ask.

The only reason I didn't speak these words aloud was out of respect for BFREE. After all, I was representing not just myself but BFREE, as well.

She could no more wrap her head around my lack of an advanced degree, than I could around her offensive and ostentatious question. She had the worst case of PhD-itis I have ever encountered. Where's a tall pair of rubber boots when you need them most?

# A National Geographic Expedition

## *Harpy Eagles and Gastropods*

Any biologist, worth his, hers, and anything in between, weight in snails, daydreams of being a lead investigator on a National Geographic expedition. And here I was, about to actually do just that! Remember I said we were tired of watching the Nat Geo channel, we wanted to go live it? I never imagined it would be land snails that got me there!

The expedition would be divided into four teams: The Harpy Eagle/ Vegetation Team, The Land Snail/Sinkhole Team, The Drone Team and the Support Team. Each one would play a vital role in the 10-day expedition deep into the jungle. It would be the most extensive and largest expedition ever launched into the Bladen Nature Reserve, the culmination of years of avian and mollusk study and funded by National Geographic.

The equipment and food alone would require ten porters from surrounding villages carrying heavy backpacks loaded with lab equipment, climbing ropes, portable generators, and enough food to feed an army for 10 days. While little more than 8 hours from BFREE, the destination is a difficult walk (some of it wading waist-high in the Bladen River) without all that gear but with heavy paraphernalia, it's just plain insane! But we were determined scientists and with careful planning, the quest could be accomplished or so we thought. But as the old saying goes, "The best laid plans of mice and men....." There would be a few unforeseen and painful glitches.

Five years earlier, while on a bat expedition in the Bladen Nature Reserve in Quebrada de Oro, the Lost Valley of Gold, Tom Pop and I decided to check out the hills above the basin, mainly looking for—what else?—lands snails—when we stumbled upon a really deep hole around 300 feet to the bottom. This thing was massive, big enough in diameter to fly a small helicopter into. With cycads growing on its vertical walls, it was a miniature lost world of tiny palm trees at the bottom. The mysteries of what might be found at the floor of this massive sinkhole (one of hundreds scattered across the Belizean landscape) would haunt me for the next five years.

A half decade would pass before I saw that crazy big hole again. We would return with a small army to scratch for land snails and listen and look for harpy eagles.

You are doubt wondering what do land snails and harpy eagles have in common? A great question. Very little in terms of physiology but in terms of intermediary organisms as part of the food web, a whole lot more than any of us had realized.

The purpose of the expedition would be threefold: to investigate and document the presence or absence of any adult or juvenile Harpy Eagles in the area where they had been documented off and on for the past ten years as well as abundance of prey items; to determine if there was any correlation between land snail species diversity or abundance and the presence of Harpy Eagles with a bottom up approach from snails to eagles; and to explore the mysterious sinkhole for new species using the latest technology, drones equipped with Lidar capable of thermal imaging. Sounds simple enough!

But this was no ordinary expedition team. My co-investigator, Jamie Rotenberg, a professor at University of North Carolina at Wilmington, and the lead investigator of the decade long Avian Conservation Monitoring and Harpy Eagle Project with BFREE, put together a talented team capable of the challenges that lay ahead.

The Snail/Sinkhole Team would consist of two elite deep-cavers all the way from Poland, Kasia and Pawel. This dynamic duo were beasts when it came to the level of expertise in cave exploration. Kasia, routinely spent days on end in caves miles deep into the Earth and Pawel's rope expertise was honed on oil rigs in the Arctic. They were indeed a force to be reckoned with! Along with these two outstanding cavers, we were joined by my nephew and niece, Aaron and Amy, a husband and wife team whose extensive rock climbing experience, along with their youth, rounded out the sinkhole team.

In order to drop this massive unknown sinkhole, Kasia and Pawel oversaw the several days of training and safety reviews on massive ceiba trees at base camp before certifying anyone who wanted to drop into the abyss.

The Land Snail/Sinkhole Team also contained Judy, several Belizean technicians, and me but each team also coordinated and assisted each other during the whole research process. The Harpy Eagle Team would be led by harpy eagle experts Jamie Rotenberg and William Garcia, along with several grad students from University of North Carolina at Wilmington.

The Drone Team contained individuals from the University of California San Diego who had just completed ground-breaking Lidar surveys in Guatemala, uncovering literally hundreds of unexcavated Mayan ruins using the most advanced technology, and here they were with us on our expedition!

Our research did indeed support an interesting connection between the presence of harpy eagles and the abundance of land snails. The mutual link would include two familiar animals, the common opossum and white-nosed coatimundi or "quash" as it's known to Belizeans. These two mammals are of major importance to the harpy as a primary food source. And guess what? Land snails were a significant food source to these two mammals.

We recorded 61 species of land snails, around 1500 specimens, within a search area of 0.03ha of forest. The Common Crater was the most frequent species detected (172 individuals). Although we did not find significant differences in species richness among our three stratified zones (flat, hillside and ridge top), we did find species diversity greatest in the flat, followed by the top, and lowest numbers of snails on the hillside. Our data compared well to other known tropical locations where snail studies have taken place, such as Borneo and Cameroon, Africa. Extrapolating from the number of individuals found within our quadrats, our plots in Belize have a significantly greater number of individuals and biomass per hectare than either of these two tropical locations. Ok enough boring science-based bullshit on land snails!

We recorded 206 bird species while inside the BNR. This constitutes 67% of the known avian diversity of the Reserve. Included in our survey were many notable species for Belize including: Harpy Eagle, Solitary Eagle, Tody Motmot, Keel-billed Motmot, Lovely Cotinga, and Tawny-throated Leaftosser. Other bird species observed and suspected consumers of land snails included large turkey size birds such as the Great Curassow and Crested Guan, also harpy food.

The expedition was hot, muggy, tick-infested, dripping with snakes, tapirs, and jaguars. You know—paradise—if you are a half crazy biologist from the backwoods of Kentucky. That said, while the jungle is no less than amazing with its endless wildlife experience, it is not without a physical and mental cost. The price you pay is not so much in dollars but in bodily assaults from flora and fauna. Remember everything out here is either sticking, stinging, or biting you.

One such unfortunate wildlife encounter was suffered by Judy. She was the unintended recipient of numerous "ticking tick-bombs" that covered her with not a few, but hundreds of tiny seed ticks, a severe tick infestation that nearly put her in the hospital.

On day two of our adventure, the Snail/Sinkhole team embarked on it's second trip to the massive abyss, having carved and flagged a rudimentary trail the previous day. It was now time to follow the flagging put there as a precaution in case the team's mission lasted into the night. Surprisingly to me, Judy boldly asserted herself to be the lead hiker. My previous experience on numerous expeditions into the Bladen and my knowledge of Judy's adverse bodily reactions to the bites of ticks started sending off alarm bells in my head. "*What was she thinking*?" But that's just it! She wasn't. Caught up in the moment and sheer excitement of being a part of something most only watch on TV, not to mention the presence of our niece and a few other women on the team to set the bar for ladies high, she plowed ahead.

I gave her that "*Are you sure you want to do this?*" look which was received like a lead balloon. My reluctance to speak up, anticipating a seriously negative outcome, would haunt me for months. I should have been able to voice my concern and give her some balance in the heat of the moment that would have produced a different outcome. But I didn't. Not because I was afraid to articulate the likely outcome of a tick infestation of the worst kind, but I didn't want to appear to be just a another controlling husband who didn't value my wife's decisions. How impractical of me that was!!

So off she trampled up the trail of pride, every step putting her closer and closer to regrettable misery. At the end, Judy would be covered from head to toe including everything in between by an endless parade of blood-suckers. We spent days pulling ticks from every inch of her body. The only place ticks didn't dig in were fingernails and toenails. Even though we removed every single tick we could find, she would later develop tick fever and many of the bites became infected, forming dozens of red streaks.

To her credit, Judy tried desperately to continue on with the work at hand. Not physically or mentally able to deal with any more biting insects, she spent the last few days, zipped inside a screened tent perched next to our makeshift lab, dutifully recording data as the snail team and I picked through leaf litter to identify specimens.

Fearing that the tick fever was leading to sepsis based on the red streaks emanating from some of her scabs, we made the decision to depart a day early to seek antibiotics to calm the raging infection within her body. But returning to BFREE was at least an eight-hour walk. I was not sure she would be able to make it on her own, so we asked Tom Pop, our good friend to join us, just in case I had to carry her out.

Upon reaching BFREE, it was a six-mile journey on the road of lost soles to the main highway, and another forty minutes to the nearest medical clinic which was closed due to it being Easter Sunday. But in Belize doctors are more accessible than in the states. So the life-saving medicine was but a doorbell away. This is the way of the jungle, *where there is no doctor.*

The expedition produced some valuable information on how and why harpy eagles are able to survive on a different but sustaining diet in Belize where there are no sloths (the main diet of the bird farther south in Panama). In other words, the Bladen Nature Reserve offers the eagles a robust ecosystem with high land snail diversity and numbers which in turn, feeds other animals that are eventually fed on by the large raptor.

And you all thought that land snails were just plain uninteresting creatures! The lesson here is simple; *"Snails rule and everything else drools"*. If you learn this simple philosophy, life will have meaning and on your death bed, you shall receive "total consciousness".

# Hippies, Yuppies, and Yippies

## Fall of the Hippies

Remember the 1970s, when it was hip to be a hippie? Ok, maybe not if you're under 40. It was a time of free-spirited individuality, unity, and love for everyone and everything. Carefree days would be spent talking about the meaning of life and running through fields of gold—not how much you made last year but how happy last year made you. A time when kids were asked what they wanted to be when they grew up they would always reply a doctor, lawyer, teacher, fireman or that oddball individual, a biologist. When kids are asked that same question today "its I want to be rich."

When we first arrived in Belize in the fall of 2006, life in this tiny Central American country was like stepping into a time machine, likened to the hippie days of the 1970s in the USA. Neighbors were more like family and parents had to send out search parties to find their kids—kids who would rather play outside all night than be inside, glued to the boob-tube (as my dad called it) or addicted to silly video games, getting ever fatter on a diet of Hot Pockets and nachos and lack of exercise. It was a time when the idea of helicopter-parenting would have been laughed at and considered as an ill-conceived and ridiculous idea.

Belize, in most places, is still a place where 5-year old kids walk to school without guardians or carry two-foot long machetes to school to help the teacher cut grass around the playground so that they can play football (soccer).

Can you imagine this scenario in the US today? A kid caught with a nail clipper is in deep shit and parents certainly would be turned in to social services for allowing their kid to walk to school alone.

## Rise and Lies of the Yuppies

The Hippie movement of the 1970s in the USA slowly slipped away, giving rise to the Yuppies, a generation more concerned with material objects and financial success than the all-for-one and one-for-all Hippie days. Yuppies were young, educated, business types often undeservedly wealthy, arrogant, and obnoxious with most living in or near a large city. This narcissistic bunch of upwardly-mobile professionals, often the children of doctors and lawyers with Master's degrees from Ivy League universities, would not have been caught dead living in a house under 5000-square-feet with less than 5 bathrooms. You know, the kind of house that would accommodate at least 10 Mayan Families and their livestock.

While most of the mere mortals of the middle class could only look on with envy, the focus of an entire generation became "bigger is better" and "He who has the most toys, wins". This erroneous thinking led many to psychiatrist couches or worse when they failed to live up to the expectations placed on them by mass media.

The effects of the Yuppie movement are still entrenched in many sections of the USA, even today. Many Americans are little more than carbon copies of each other living in big houses on a hillside that are virtually indistinguishable from each other. The bigger the houses became, the more money was needed to pay for them. And when you have space, guess what? You fill it with stuff. The more stuff that was accumulated, the more discontentment followed.

But Yuppies have been sold a faulty bill of goods by corporations seeking to expand and line their own pockets with cold hard cash while giving little regard to the impact their pandering would have on the very fabric of society and the happiness of its members.

**Awakening of the Yippies**

There's Hippies and their polar opposites, the Yuppies and then there's the up-and-coming Yippies, a new movement taking hold within the current generation Z. A kind of hybrid model *Homo* that combines the best of each of the previous trends, leaving the worst components of each behind.

While most people we met in Belize live in the moment, many in the US waste precious time fretting about the past or frantically planning for the future. But the past is gone and the future is certainly not guaranteed, there's only the moment where most Belizeans live. That's what the Yippie movement is all about, a kind of reboot of those Hippie days. Buying less shit means less money needed to live and more time to play, spend with your kids, family, and friends or travel and see the world.

If it's true that all humans share a common goal which is simply "happiness", what are the primary factors that get us there? Why does happiness come so easily for the poor, while remaining allusive to the rich? The answer lies in the strength of social interactions.

A 2014 Harvard Study concluded there was one characteristic that distinguished the happiest 10% of humans from everybody else—the strength of their social relationships. Social support was a far greater predictor of happiness than any other factor—more than GPA, a big house, fancy car, family income, SAT scores, age, gender, or race. So human wellbeing is best served not by our relationships with possessions, but by our relationships with each other and let's not forget our life-giving Mother Earth. And guess where some of the happiest people live? You guessed it, BELIZE! The US doesn't even make the top 10!

That first year and our subsequent return to the USA for a visit would reveal stark comparisons of lifestyle and priorities that had eroded right under our American noses, the subtle changes happening gradually over time. It was an eye-opening event that further exposed this disparity, where the amount of money we make and power we wield has become more important than the children we shelter.

Leaving a country filled with some of the happiest people we had ever met, we were about to enter a world where money seemed to rule nearly every life decision. We entered our country of birth and residency feeling like foreigners in our own land. A land where children are raised not by their parents but strangers, where kids are becoming increasingly overweight and, for the first time in our nation's history, expected to have shorter life spans than their parents.

In contrast, family, friends, and community play a significant role in the daily lives of Belizeans. Despite having few material possessions when compared to the typical American family, they were and still are, friendly and cheerful people.

Our first steps onto US soil would leave us disillusioned and dumbfounded for the next few months as we drove home from Miami to Kentucky. The people we met along the way at gas stations, stores, and restaurants were our family of Americans, yet they were like zombies going through the motions of life without emotion and completely oblivious to their surroundings—our presence barely acknowledged by the walking dead attached to cell phones like politicians are to votes. As we drove through subdivisions, we wondered, "Where are the children?"

Kids no longer ran and played among green grass, trees, and singing birds but sat in front of brain-numbing screens playing ridiculous video games. Adults became disillusioned and unhappy with their lives, their jobs, their spouses, and bored with their endless piles of shit. Shit that spilled over from jumbo houses into endless rows of rented storage spaces. Once a nation of fearless people, we have become a nation afraid of our own shadow—anxious, sue-happy, worldly snobs living quiet lives of desperation!

Had we missed something while living in Belize for the past 7 years? As it turns out, we had, and as profound as it was, the changes were so subtle we had paid little attention to the transformation until we went to *live where the cats eat the dogs*. The shift was slower than two copulating land snails and went unnoticed as an impromptu migration of people moved from their backyard oases to their castles of plaster. It didn't happen overnight, but over decades. Our once valuable and consequential lives snatched from beneath our very noses without a trace. The lie that bigger and more is better was only an illusion propagated by the powers that be.

Perhaps the Yippies are finally getting it right, living in tiny houses, not all the same, seeking jobs of passion, not jobs of money, working to live not living to work—*living a life of less, to live a life of more.* Is this the Belizean way? You better Belize it!!

**Lost Souls of the Maya**

You're no doubt thinking, what the hell does religion have to do with *living where the cats eat the dogs*? It's the prologue to my next book entitled *"If I'm Going to Hell, At Least I'll Get there Honestly"*. Just kidding, mostly. It's quite simple, really. Religious dogma has had a profound effect on many of our Mayan friends in Belize, bringing the topic to an ever-bulging boil that needs lancing.

When it comes to religion, the Maya people remain trapped between two worlds—much of their past culture destroyed by raiding Spanish Conquistadors and self-righteous priests who denied their homegrown beliefs. For example, thousands of Maya codices inscribed on deerskins, idols, and altars, as well as story paintings and hieroglyphs immortalized on magnificent structures, were destroyed by Spanish priest Father Diego de Landa, who described these works of art and culture as nothing more than works of the Devil, designed by the evil one to fool the Maya, preventing them from accepting Christianity.

He wrote: *"We found great numbers of books (written in the characters of the Indians) but as they contained nothing but superstitions and falsehoods of the devil* [sounds like they're describing the Christian Bible to me]; *we burned them all, which the natives took most grievously, and which gave them great pain.* Reading this made me want to throw up, not because I agreed with the Mayan religion any more than I do Christianity but because I could empathize with how much this must have hurt them! These priests or so-called "missionaries" had no more right to destroy an entire culture and its belief system than I would to burn a Catholic Church to the ground.

Father de Landa was not alone in his ruthless efforts to rob centuries of tradition from the peoples of Central and South America either. A bishop from Mexico bragged about destroying 20,000 idols and 500 Indian temples, even going so far as to burn a Christian-converted Indian at the stake who "backslid" when it was reported he had reverted to worshipping a rain-God in a desperate attempt to alleviate a serious drought situation in his community.

Yet the self-proclaimed bearers of truth live on to this very hour causing havoc across the globe. Not only do they spread unsubstantiated doctrine, priests and missionaries continue to spread disease that has killed thousands of indigenous people in places like the Amazon. They also undermine traditional cultures and social cohesion by forbidding certain practices like the "**Mayan Deer Dance**" and healing with herbal medicines as things of the Devil. Ironically, 40% of the pharmaceuticals used today come from plant-based sources and knowledge gained from the wisdom of indigenous people.

Today, the little bits of Mayan culture that remain are snuffed out by Bible-thumping Christians that feed off religious uncertainty. Although they come with good intentions, Christians continue to loot the Mayan of their unique cultures who now have little connection with their past. A result of high priests and missionaries condemning them for practicing their ancient religion tied to stone stellae they believed to be their protectors. Sounds like the Catholic Church I grew up in.

The Maya are now a people caught between their denied past, lost culture, and the so-called *real truth*—Christianity. But truth is hard to come by. Perhaps it is found with one of 4300 other religions scattered across the planet such as Islam, Hinduism, Buddhism, Sikhism, Taoism, Judaism, Confucianism, Bahá'í, Shinto, Jainism, Zoroastrianism, and or the Maya. They all can't be right but they can certainly all be wrong! In reality, *family and geography are the two main determiners for religious beliefs,* in other words—inherited conviction.

The conflicts and confusion begin when different denominations or faiths bring their specific doctrines and spread dissimilar beliefs around like a 95-cent hooker, leaving a shitload of religious confusion that would dumbfound Albert Einstein. What's left in their wakes are tidal waves of clashing ideas and boatloads of confused followers which are sure to capsize. After years of indoctrination, missionaries return home, patting themselves on the back for a job well-done, never realizing the trail of tears left behind—the once-united Maya communities now estranged by newly emerging and conflicting religious views.

We have many Maya friends that no longer speak to their family members or friends as a result of religious tenets shoved down their throats by outsiders. I have grown to hate this kind of religious dogma—I'm right and you're wrong—attitudes that can't be proven any more than the tooth-fairy. They are just opinions lacking any real evidence and like assholes, everybody's got one. In their "fecal" attempt to fix a shattered world, they only leave it a little more broken. This is the downside of religious dogma that few ever talk about—the perpetual divide cultivated in family, friends, and countries.

Unfortunately, this is all done in the name of Jesus and a 2500-year -old archaic text—a book used to control a bizarre narrative that seeks to manipulate all thought that doesn't conform to its storyline as a sin punishable by an eternity of burning in Hell. Infinite torment for a finite crime. God's distorted sense of justice, I guess!

In my opinion, the worst Biblical idea ever conceived was **Heaven** and **Hell**, scaring kids half to death and manipulating the minds of adults through fear of the unknown. This indecent proposal is like God holding a gun (hell) to your head and saying, "If you don't do what I tell you, I'm going to pull the trigger. But it's your choice—after all, I gave you free will." The outlandish idea that we will burn in hell for lack of belief in a GOD, clearly quells the idea of free will.

This concept has done nothing but anchor our minds in a never-ending sea of uncertainty—"why did I take this action or inaction and what are the biblical consequences?" In other words, did I do the right thing for the wrong reason, a milkbone (Heaven) or out of fear (Hell)? Both could be viewed as self-indulgent reasons. I consider these motivations for doing the right thing, not at all, how I want to mature my intellectual self. In a letter written by Albert Einstein to philosopher Eric Gutkind in January 1954, he had this to say: "*The word God is for me nothing more than the expression and product of human weaknesses, the Bible a collection of honorable but still primitive legends which are nevertheless pretty childish. No interpretation, no matter how subtle, can change this for me.*"

We are born, we live, we die—a reality for all life forms. Why do we "think" there's something more? Because we can, not because there is! While self awareness fosters freedom of thought, it also leads to abandonment of reason.

We are not a species that has lost our minds,
we are a species lost in our minds.

This amazing image of a jaguar was taken in Belize by my good friend and film-maker Carol Farneti Foster.

# Jungle Chronicles

The following is a series of short anecdotes, highlighting the remarkable wildlife I encountered while *living where the cats eat the dogs.* It is by no means a complete account, which could easily fill another book or two. But for me these were the attention-grabbers of jungle living. While most daytime animal experiences are nothing less than amazing, nighttime is when the jungle really comes alive, and as such, where I spent much of my time alone.

There's nothing more uncanny than a moonless night under a canopy of endless green, yet more rewarding than a close encounter with a jaguar, the smells of night-flowering lianas, or the subtle sounds of bat wings as they flutter by.

### ~*Mountains of the Maya*~

*This place called the Maya Mountains is one of nature's finest moments in time, a concert of biodiversity a million years in the making. In awe I stand where spectacular rainforests shelter a profusion of life and endless waters flow past me to the sea. A place where time seems unhurried by the world beyond. Boundless sensations surround my every footstep. I am humbled by the calls of howlers but no less overwhelmed by the passing of a Blue Morpho. My senses utterly occupied with these surroundings, I will forever cherish my moments in the grandeur of the Mountains of the Maya.*

## Animals Echo Our True Nature

**Human Supremacy Over the Earth is a Myth**

As a wildlife biologist, I have learned a great many things about *Homo sapiens,* the animal, by studying the other animals with whom we share the planet. Our supremacy over other creatures is a myth that our thinking minds have created. While religions can, at best, make only unsubstantiated claims about our dominance, nature reveals unadulterated truths about our reality that should humble us all.

The latest Corona virus is a good example of what really runs the planet—microscopic parasites that invade other organisms with only one mindless goal—replication—having little or no regard to their hosts. We are no better than anything else and only our ego tells us differently!

The belief that we are the most evolved creatures on Earth is also blatantly false. Why? Because the evolution of a bird's wing is just as extraordinary as our human brain. We may be able to design a machine that allows us to defy the laws of gravity, but still remain flightless without such a device. So, each and every species is on an equivalent plane of evolution and as such, reaches its unique position through a series of evolutionary tweaks, all shaped by an ever-changing planet. This does not mean that evolution has stalled, but continues on with time, fine-tuning each organism to fit its perpetually changing environs better and better, until it doesn't. The end result? Extinction.

While most organisms are constantly being adjusted through the process of "natural selection" via genetic-mutations or alterations in genes (which can have either positive and or negative effects), some creatures have changed little as a result of reaching near perfection. Crocodiles are one of evolution's greatest achievements in flawlessness; so well-crafted that in 200 million years, there's been little change in design, outliving the dinosaurs by some 65 million years.

If crocodiles represent examples of nearly-flawless evolution, then Tardigrades are the embodiment of evolutionary perfection. Also known as a waterbear or moss piglet, this organism, in terms of its ability to thrive on an ever-changing planet, is measured not in centuries but in millenniums. Tougher than Superman, these tiny critters have survived mass extinction events like the dinosaurs, can survive temperatures of 300 degrees Fahrenheit for at least a few minutes, and thrive in temperatures of -328 degrees Fahrenheit for days! Tardigrades can live for more than a 100 years in a cryptobiotic state. Still not impressed? Waterbears can withstand 1,000 times more radiation than other animals, 6 times the water pressure present at the bottom of Mariana Ocean Trench and repair its own DNA. They have even survived the vacuum of space for at least 10 days with no ill-effects. Around for more than 600 million years, compared to our pitiful 300,000 years, Tardigrades will no doubt be here long after *Homo siphons* disappear.

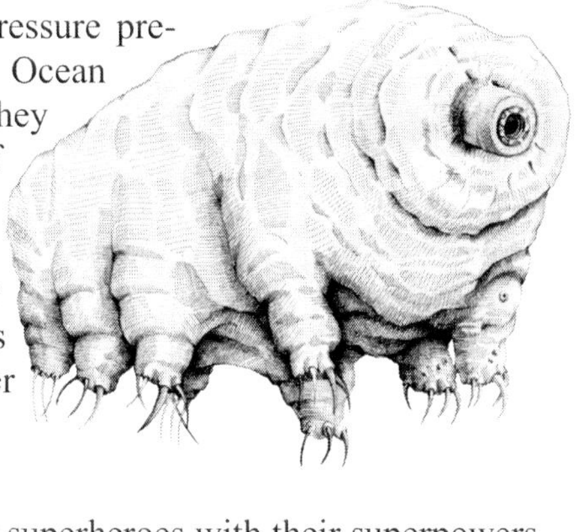

While we marvel at our comic book superheroes with their superpowers of make-believe films, everyday animals actually have many of these same enhanced capabilities—the lowly Tardigrade being an extraordinary example. We may carry the most evolved cognizant brains of any animal, but clearly are not the strongest, fastest, longest-lived, or even the best designed life forms on this planet. So one animal is not more evolved over another, but rather differently evolved to fit its particular environmental requirements. There are, of course, ever evolving exceptions to the rule. And its these exceptions that either favor or hinder a species' particular needs in an perpetually-changing world. This is how and why the Earth has so many plants and animals today and why the term "normal" is often an unfamiliar reality in nature.

## What's Normal

While gender and sexual expression in many organisms does not align with our human ideas of what is considered "normal", it may in fact be more aligned than we think. As much as 5% of some species of ducks and geese—known to partner for life—are neither bisexual or heterosexual but completely homosexual, as are some humans. Fish can change sexes overnight and hermaphrodites such as land snails, are both female and male. Some animals can even screw themselves and produce offspring. We wrongly assume that sex is always between same species of the opposite sex, resulting in progeny. We also mistakenly think that homosexuality is an unnatural behavior among our species. But in the animal kingdom (of which we are a part), homosexuality, along with many other human taboos, are often the everyday norms.

**Two Female Bonobos pleasuring each other**

Alan Dixson

Bonobos, for example, are entirely bisexual; both males and females engaging in hetero and homosexual behavior, and are the most promiscuous of any non-human species. In addition, up to 75% of bonobo sexual behavior is purely for pleasure. Female with female is especially well-documented, in particular with around 60% of bonobo sexual activity occurring between two or more females—female to female ventro-ventral mounting, with genito-genital (GG) rubbing (illustrated above). Furthermore, sex partnering between juveniles and adult bonobos is not uncommon. While homosexual bonding in bonobos represents the highest frequency of homosexuality known in any primate species, homosexuality has also been reported for all the great apes including, chimpanzees, gorillas, orangutans, and, of course, humans. Interestingly, Dutch primatologist Frans de Waal believed that the increased sexual activity between bonobos might be their way of avoiding detrimental conflicts. Sounds good to me!

Believe it or not, over fifteen hundred other species of animals demonstrate same-sex attraction or homosexual behavior. Animals also have sexual fetishes, masturbate (sexually-frustrated male turkeys have been observed screwing cow pies for a lack of a partner), partake in fellatio (female Short-nosed Bats give head-jobs to males), have sex with inanimate objects (orangutans make and use dildos of wood and bark), practice cross-species sex (bestiality), sexual cannibalism (the female praying mantis will eat her mate after copulation), sexual prostitution, and even necrophilia (having sex with the dead). Sexually frustrated male kakapo parrots of New Zealand have been filmed mating with dead seagulls. Talk about a double taboo!

Hormonal and environmental factors which are in a constant state of flux can, and does, drive a diversity of reproductive strategies and sexual partner selection. As humans, we like to categorize our behaviors as either normal or abnormal, but a closer look at some of the animal kingdom and our closest primate relatives tells a very different story and begs the question, "what's normal"?

Another example of animals echoing our true nature can be found in adolescence. We were recently given a baby raccoon named Roscoe from my daughter that a neighbor rescued from a fallen tree. Baby raccoons are like puppies in terms of their sense of wonder and playfulness and Roscoe was no different. She often slept with Judy and we had a most gratifying time raising her from a youngster. But our intention from the beginning was to re-wild Roscoe back into nature. So we took her on morning walks where I taught her how to harvest wild berries and fine and eat crawdads under rocks in nearby streams. Personally, eating raw crayfish is like chewing eggshells—there much better cooked!

As time went on Roscoe grew out of her playful and carefree days becoming more serious with each passing month. Aside from otters which seem to never grow out of there playful youth, few animals break free from the fetters of ever deepening and stern behavior. And while there's an uncanny parallel between the wonder and mischievousness of youth shared by both animals and humans alike, it doesn't end there. Analogous behaviors can also be detected in both groups as they mature into adulthood. As rambunctious energy wanes, a serious side naturally begins to surface. For most animals, life becomes more acutely solemn, eat or be eaten—not so different from human societies!

Even though our animal side in both youth and adulthood may vary little from our animal cousins, our modern lifestyle and cognitive brain allows us to rewire some basic instinct that has enslaved most other species to just basic survival. We are no longer strapped by the labors of hunting, gathering food or hiding from predators. Consequently we can reshape our trajectory, choosing instead a different evolution of one's self—clinging to that lifeboat of youth and wonder we so loved as kids. Embedding ourselves deeper into humanity's insanity only moves us further from natures wisdom of youth and its benefits.

***Our greatest nemesis will always be our thinking brains.***

## Natures Uncanny Secret~ SYMBIOSIS

In many ways, the whole Earth is a kind of symbiosis with one notable crack, *Homo sapiens*. Before this creature arrived on the scene, the planet functioned more or less as a unit or Gaia; an ever-changing, but self-correcting, life supporting system. According to a 2020 paper in *Science Advances,* there have been at least six mass extinctions, wiping out 99% of all life that ever lived, and yet, there is an astonishing number of species living today. While the 7th mass extinction is well underway, a result of actions or inactions taken by humans, Earth does in fact have an expiration date in its future from an inescapable heat death, which can be described as: *That ultimate state of thermal equilibrium implying conditions of maximum entropy and zero available energy that according to the laws of thermodynamics, the material universe is apparently approaching.*

But not to worry, we will no doubt remove ourselves from the equation long before that unfortunate day. Have you ever heard that saying "We are too smart for our own good?" It appears to be true when considering our collective ability to push Earth to premature death. It's no Hollywood movie with a happy ending, but a foreseeable reality all tied to that one mixed blessing, the evolution of our brains. With increasing intelligence came knowledge and discovery, but not without its problems. Higher intelligence gave rise to *self-awareness* and one of our greatest single liabilities—*arrogance*. It's what really separates us from the animals. If Ernst Mayr was correct in his contentious idea that human intelligence was a lethal mutation, are we all just destined to fail ourselves and each other? Can we ever achieve true symbiotic balance with the Earth?

For humans, its become a lop-sided-relationship with money, stuff and power. A materialistic parasitic feeding on Earth's finite resources like cancer feeds on healthy cells. While a Symbiotic-Relationship might get us to an attainable future, at least for a while longer, the former is driving us toward annihilation. Earth fosters our physical wellbeing, and in return, we should protect Earth—by any means necessary.

Clearly, human security is underpinned by our relationships with a healthy thriving planet. A bond that should never, under any circumstances, be separated!

## It's Simply Evolution

The driving forces behind evolution can be summed up as allele frequencies or changes in populations over time. Natural selection, genetic drift, mutation, and gene flow are the mechanisms that push forward these changes. When one or more of these forces are acting in a population, evolution occurs. Put in more basic terms, if a genetic mutation gives a species an advantage in its surrounding environment, it's more likely to flourish and replicate—evolution—but if that same mutation is an obstacle to survival, the result may be extinction. Take, for example, a gene mutation resulting in a black moth. A darker moth is less likely to be seen and eaten in an environment with darker backgrounds, but more likely to be seen and eaten in a lighter background. Therefore, a black moth in a darker environment will survive to breed, passing on its advantageous darker gene to its offspring, while a white moth in the same dark environment will not likely survive and multiply.

## Evolution, Abiogenesis, and Something from Nothing

Many theists will argue that evolution is just a theory, not a fact. Actually, a "theory" is the graduation of an idea or hypothesis, having the highest degree of certainty that science can declare. A hypothesis, on the other hand, is an idea that hasn't been proven yet. If enough evidence accumulates to support a hypothesis, it moves to the next level known as a theory and becomes accepted as a valid explanation of a phenomenon. In other words, the "theory of evolution" is a scientifically tested and well-proven fact, while a God claim is still just a weak hypothesis, according to science that is.

Evolution is the best scientific description of the process by which life has evolved. Several lines of evidence for evolution include **intermediate species** found in fossil records, **vestigial traits** (parts no longer used) such as non-functioning eyes in blind cavefish, the hip bones on whales, and leg bones buried in the muscles of some snakes, **biogeography** predictions made by evolution, and the best confirmation of evolution to date, **genetic evidence** which puts common ancestry beyond a reasonable doubt. It is generally accepted that chimpanzees and bonobos are human's closest living relatives, with both species sharing around 98.7% of our DNA, as compared to around 50% humans share with a banana.

**Biogenesis** is the hypothesis (not a fact) that living things can only come from other living things through reproduction. **Abiogenesis** is the hypothesis (not a fact) that life arose more than 4.5 billion years ago on Earth from non-living matter such as simple organic compounds that contained carbon-hydrogen bonds. The first life forms generated would have been simple and gradual, becoming increasingly complex through the progression of evolution. **Panspermia** or as I like to call it, "Universal Ejaculation", is the hypothesis (not a fact) that life exists throughout the Cosmos and is scattered by meteoroids, asteroids, comets, space dust, and also by spacecraft.

One burning question theists often ask atheists is, "How can something come from nothing"? My answer is simple. I have no idea, and more importantly, don't really care. Arguing about how it all began 12 billion years ago is like two turkey vultures arguing over which parts of rotting meat to eat first. In the end, it all comes out looking the same! Perhaps there was always something, rather than nothing, although nothing implies something rather than nothing. In other words, the cosmos, quite possibly, had no beginning and always existed in some form or another. That's the hypothesis that makes the most sense to me.

**Earth's Greatest Triumphs and Worst Failures in Evolution?**

# Belize, a Biological Wonderland

**The Monkey Bridge**

Central America, including Belize, is a mixing pot of both temperate wildlife (i.e. white-tailed deer and bear) and tropical animals (i.e. monkeys, anteaters, opossums and porcupines). This is a result of its unique geographic position connecting the two great continents of North and South America, often referred to as the "**Monkey Bridge**".

Undoubtedly one of the highlights of a Belizean rainforest visit is the haunting grunts and yowls of the howler monkey. This medium-sized primate is world-renowned for its ability to produce high-volume reverberations that can clatter the brain, sounding a lot like being in the middle of Jurassic Park. Reported as one of the loudest land mammals on Earth, Yucatan black howler monkeys are endemic to the Yucatan region of Mexico, eastern Guatemala, and Belize. Forest fragmentation is considered the single largest threat to the survival of this species and, in many areas of their former range, this distinctive primate has all but vanished.

At BFREE, we shared our home, let me restate that—the howlers shared their home—with us for seven wonderful but vociferous years. These amazing primates are easygoing, leaf-eating monkeys while Central American spider monkeys are hyper-energetic, due to their high glucose intake of native fruits.

Like howler monkeys, spider monkeys have disappeared over much of their former range, however, the species remains a familiar face in the Bladen Nature Reserve. Large troops of forty or more have been observed, although smaller groups of around a dozen are most common. With long arms and a prehensile tail to match, it is easy to see where these primates get their name.

Spider monkeys are the acrobats of the canopy, leaping from branch to vine to tree with amazing agility. Interestingly, spider monkeys not acclimated to humans, put on an amusing performance, breaking branches, urinating, and even defecating, all of which is directed at hitting any potential threat square in the face. If monkeyshines fail to drive the intruder away, they simply lose interest and disappear into the jungle. We would come to love and, sorely miss, the antics of spider monkeys and the ear-splitting calls of howlers.

**Ocelot**

## Cats of Belize

The Bladen Nature Reserve is home to a remarkable number of both familiar and uncommon animal species. The exceptional diversity and abundance of wildlife are a result of its geographic location in Central America as well as the geological and vegetative complexity of the landscape. No roads and only a few scant researcher trails penetrate the deep wilderness. The pristine integrity of the Bladen Basin allows animals to move freely. Unimpeded by human development, it supports a large number of cats in a relatively small area.

Pinnacle feline predators of the Maya Mountains include five species: the jaguar, ocelot, and margay—all three cats strikingly spotted—along with the puma and jaguarundi—cats with no spots. The jaguar is the largest of the five cats, weighing upwards of 200 pounds and are formidable hunters found in deep jungle environments. The big cats stalk a variety of prey including armadillos, deer, pacas, peccaries, coatis, and tamanduas. Reptiles such as green iguanas and turtles are also taken.

Rare melanistic, black color-phases of jaguars are sometimes seen in Belize. Pumas may occasionally possess these black color morphs as well, but, in general, are extremely rare.

The most common cat in the Maya Mountains is the ocelot—a medium-size feline weighing in at around 22 pounds. Principal prey items in Belize include armadillos, opossums and reptiles especially snakes. In the 7 years I spent in Belize I observed well over a 100 individuals, mostly during night walks. The margay, at only eight pounds, is notable smaller about the size of a hefty house cat. Margays are the most arboreal of the New World cats and can rotate their hind feet 180°, allowing them to run down the trunk of a tree head first like a squirrel. Margays eat reptiles, squirrels, opossums, climbing rats, and mice.

The non-spotted cats of the Bladen Nature Reserve are the jaguarundi, around 15 pounds, and the much larger puma, weighing in at around 100 pounds. The puma, also known as mountain lion, cougar, panther, and in Belize, red tiger, is Americas most wide-ranging feline. It can be found from middle Canada to southern Chile in lowland forests and deserts to over 5000 meters. Although seldom seen, pumas are relatively common, favoring medium to large prey (deer, gibnuts, and agoutis). Uneaten portions of their food are often covered with sticks and other forest debris and marked with strong-smelling urine.

**Vintage illustration of a jaguarundi**

Looking somewhat like an otter, the jaguarundi is only slightly larger than the margay but smaller than an ocelot. Jaguarundis may be entirely reddish or dark gray, the latter being the more common color phase. In Belize, three radio-tracked individuals traveled about 6 km a day and occupied large home ranges. The species favors areas with low dense cover near water and is less common in deep jungle. Diet varies to include arthropods, small rodents, birds, and lizards.

**Mountain Cows, Pigmy Deer & Smelly Swine**

The Central American tapir (pictured here with track) is locally known as the mountain cow. It is a large semi-aquatic mammal weighing up to 600 pounds that is related to the rhinoceros, and although rare elsewhere in Central America, a common resident of the Maya Mountains and Bladen Nature Reserve. At home in water or on land, the vegetarian tapir has poor eyesight, but a keen sense of smell. Its large three-toed tracks are a common sight.

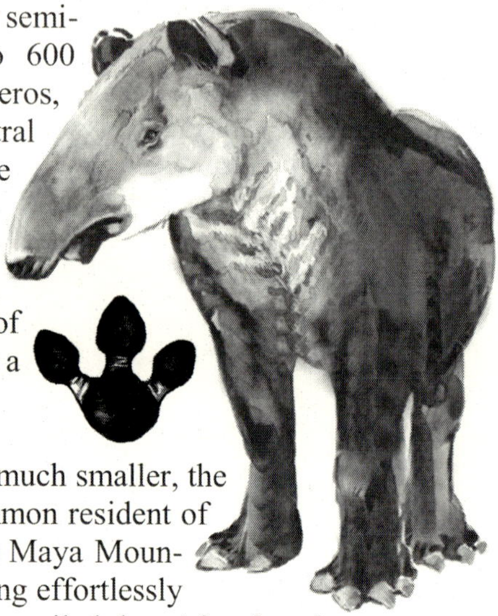

Similar to the white-tailed deer but much smaller, the dog-sized red brocket deer is a common resident of the Bladen Nature Reserve and the Maya Mountains. This is a jungle species, moving effortlessly through the dense jungles. The white-tailed deer (also found in Belize) is a species of savannas and floodplains.

Great herds as much as 200 strong of the endangered white-lipped peccary, also known as warrie, along with collared peccaries, still roam the BNR. The pungent warrie uses scent glands to mark its territory, so the pigs are more often smelled than seen. Among the most feared wild animals in Belize, when threatened, warries make loud clicking sounds with their formidable canines, and have been known to attack humans. In large numbers, these wild swine are a force to reckon with!

**Weasels**

Three of the most recognizable weasels found in the Bladen Nature Reserve are the greater grison, the tayra, and the neotropical river otter. Grisons may be active day or night and rarely climb trees, eating a variety of reptiles, birds, and small mammals. Tayras are larger, mainly diurnal, sometimes crepuscular, excellent climbers, and are often seen in trees foraging for fruits, invertebrates, tree rats, and occasionally, even young monkeys.

Neotropical river otters are essentially diurnal mammals living in rivers, swamps, and other large bodies of fresh water. Burrows are dug in the banks along waterways. Diet consists primarily of fish but also includes aquatic snails, freshwater mussels, crustaceans (crabs and shrimp), small mammals, snakes, turtles, and, occasionally, birds. The best sign of otter presence is their smelly (odor of fish) scat left on protruding log or rock structures in and around the water's edge.

Of the 12 species of skunks known to science, four occur in Central America, and two (possibly three) are found in the BNR; all belonging to the family Mephitidae. Skunks are best known for their odoriferous anal secretions which can be discharged at will and directed at any threat. The smelly spray can cause nostrils to scream for mercy and eyes to burn. When skunks feel threatened, they stomp their front feet first (a warning), next facing their rear end toward the adversary and squirt away; the spray reaching more than fifteen feet in some cases. Two common residents of the BNR include the spotted skunk, a good tree climber, and the larger striped hog-nosed skunk.

## Armored Mammals

Armadillos are a bizarre group of strangely-shaped and armored mammals which include 9 genera and around 21 species, most occurring in South America. Two species are found in the BNR, the nine-banded armadillo and the much rarer naked-tailed armadillo. Armadillos are covered with a bony protective armor and roll into a ball to protect themselves from predators. This defensive behavior protects the animal's unarmored belly region against predator attacks. Armadillos use enlarged front claws to dig for ants, termites, grubs, and other invertebrates as well as excavate underground homes. Their abandoned burrows provide essential subterranean habitat for snakes, amphibians, and smaller mammals.

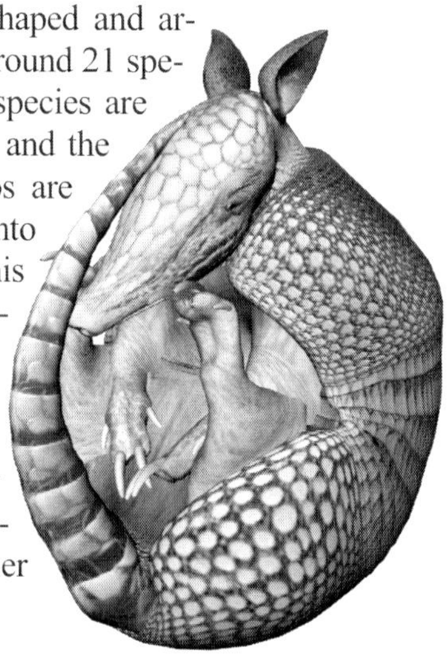

## Nightwalkers and Cacomistles

Cacomistles, like kinkajous, are arboreal mammals found in tall trees where they feed on a variety of foods including the fruits of *Ficus, Cecropia,* and *Poulsenia* trees, small vertebrates, and insects. The species is strictly nocturnal, its presence detected by loud, wailing calls heard after sunset and during the night. Cacomistles are reported to be uncommon to rare with patchy distribution in Central America and southern Mexico. In Belize, the species is relatively common in higher elevation forests of the Maya Mountains.

The kinkajou, locally known in Belize as nightwalker, is a member of the Procyonid family (raccoons and allies). With five arms, if you include a prehensile tail and oversized eyes that act more like a night-vision scope, the noisy kinkajou is more often heard than seen, jumping among the cohune palms making loud crashing noises in search of fruits, seeds, and tender leaves.

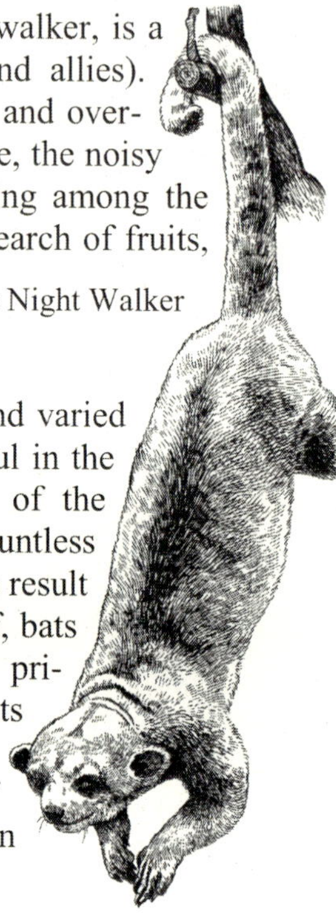

The Night Walker

## Chiroptera (Winged Mammals)

With more than 75 species, Belize has a rich and varied bat community and they are particularly plentiful in the Bladen Nature Reserve, comprising the bulk of the mammal diversity. This is due, in part, to the countless cave systems found in the BNR, but also likely a result of high tree diversity. Contrary to popular belief, bats are not flying rats but more closely related to primates, lemurs, and tree shrews than rodents. Bats are long-lived (some species living more than 30 years), social animals that possess some of the most highly-evolved parenting skills seen in mammals.

The diet of bats varies greatly in tropical regions, consisting of fruits, nectar, blood, insects, frogs, fish, birds, and even other bats. In the BNR, bats have also been recorded feeding on arboreal snails such as the princess cone, banded dome, and Belize tuba gleaned from trees and rock features. Fruit and nectar bats are considered critical seed-dispersal and pollinating agents of numerous species of trees. Without these important services, certain trees would suffer large-scale declines.

With a face that looks like it's been beaten with an ugly stick, one of Earth's most abominable-looking and yet creatively-evolved Chiroptera is the wrinkle-faced bat. This bizarre-looking bat represents an evolution that paid more attention to function than good looks. Wrinkles on the bat's face funnel juice from masticated wild figs into the protruding lower mouth for consumption, nature's hideous but effective juicer.

The amazing evolution of the lips of some Chiroptera species such as the fringe-lipped bat is thought to enable the species to differentiate poisonous frogs from non-poisonous ones while on the wing. If toxic frogs are detected, the bat simply avoids catching the frog and flies off looking for less toxic species to consume.

A frugivore, the great stripe-faced bat, eats wild fruits and is considered to be one of the most important seed dispersal animals in the jungle. The modified lip pustules are thought to help the bat sense different foods for palatability. The largest bat in the Americas, the great false vampire bat, *Vampyrum spectrum,* has a meter-wide wingspan. These large bats feed on tree rats, parrots, trogons, and even other bats. They roost in caves and trees of the Bladen Nature Reserve and are considered uncommon. One of the coolest bats is the famous tent-making bat of Central America. This small Chiroptera chews the mid-vein of *Heliconia* leaves which causes the large leaves to droop and form a protective tent.

The enlarged ears of some bat species are designed to detect the sounds of not just flying insects but crawling insects, which are promptly gleaned from vegetation while hovering like a hummingbird.

The highly-engineered fishing bat is like no other Chiroptera, having downward turning ears, enlarged back claws for snaring fish, oversized cheeks for holding the catch, and plastic-like wings that repel water.

## Pouched Mammals of the Americas

When one hears the word marsupial, those strange-looking kangaroos from Australia are usually what comes to mind. Opossums are the marsupials of the Americas and are exceptionally diverse, comprising more than 90 species, occurring mostly in Mexico, Central, and South America. Six species are found in the Bladen Nature Reserve including the water opossum, Virginia opossum (widespread in North America), common opossum, Mexican mouse opossum, Central American wooly opossum, and the gray four-eyed opossum.

With bulging red eyes, the Central American wooly opossum is one of the more colorful species. Unlike other opossums, its tail is densely haired and the species is decisively arboreal. Interestingly, wooly opossums are thought to be important pollinators for several species of trees.

Threat behaviors displayed by opossums include loud hissing sounds with sharp metallic burst calls that are startling and downright earsplitting! Opossums also use a variety of other scare tactics such as opening their mouths widely, drooling, and raising a clenched fist. When threatened by a predator, the Virginia opossum has perhaps the strangest defensive behavior of any mammal. Faced with extreme fear, the Virginia opossum reacts, not by running for its life, but by fainting, opening its mouth, drooling, and secreting smelly anal fluids—a condition referred to as "playing possum".

This behavior can be compared to fainting in humans. In both opossum and humans, fainting is an involuntary comatose-like state, resulting from a traumatic event. This defense strategy has worked to the opossum's advantage by fooling predators into thinking it is dead and its flesh is spoiled. Predators such as cats that are stimulated by the chase and prefer a fresh kill are discouraged. There is some evidence that suggests Virginia opossums purposely harass (not eat) marine toads for their protective poison. By rolling over the stressed toad, opossums are thought to rub the deadly toad toxin into their fur to ward off attacks from predators, or perhaps rid themselves of parasites such as ticks and fleas. It is not known if other opossum species exploit toads in this manner.

One of the most remarkably evolved mammals in the BNR is the water opossum, a semi-aquatic opossum possessing back feet that are webbed used for swimming on the surface and underwater; its diet consisting mainly of crustaceans (crabs and shrimp), and a variety of fish.

**Mouse Opossum**

Not all New World opossums contain pouches. The Mexican mouse opossum is without these maternity pockets and like most mammals, suckles its young externally. Most species of opossums are short-lived usually not surviving past one or two breeding seasons. All species are solitary outside mating encounters. Mostly omnivorous, opossums eat a variety of food including carrion (dead animal matter), seeds, fruits, and grasses. Opossums are outfitted with fifty teeth, the highest number recorded for any species of land mammal in Central America.

## Quills of a Different Feather

There are roughly 16 species of porcupines in the New World, mostly occurring in South America, with two species found in Central America and one species in North America. They are more closely related to pacas and agoutis than Old World (African) porcupines. The group is best known for their modified hairs called quills used in defense. Contrary to popular belief, quills are not thrown or shot but easily dislodge and anchor into the soft tissue of animals that dare harass them. The

Mexican porcupine, the only species in Belize, is largely arboreal and, like the howler monkey, has a prehensile tail used as a fifth arm while climbing trees. The species is nocturnal, sleeping during the day in hollow trees or liana tangles. They feed primarily on seeds, fruits, buds, and young leaves of *Cecropia*, *Ficus,* and *Bronsimum* trees. Although generally reclusive and quiet, Mexican porcupines produce loud yowls and screams during the breeding season.

## Small Mammals

The importance of small mammals (rats, mice, voles, gophers, moles, and shrews) in ecosystems cannot be over-emphasized. These bundles of protein are critical food sources to avian hunters (hawks, owls, and falcons) and terrestrial predators (cats, weasels, foxes, and snakes). Small mammals are also vital agents of seed dispersal in forest environments, transporting and burying millions of seeds every year, many of which are forgotten and grow into trees. Vigorous ecosystems favor healthy populations of small mammals, in turn lessening predatory pressures on animals like the great curassow.

Shrews possess venomous saliva used to subdue prey and feed on a variety of potentially damaging grubs which attack, weaken, and sometimes kill roots of trees and shrubs. By tunneling through the ground, small mammals such as gophers and moles not only aerate soils but loosen soils allowing better absorption of rainwater which greatly reduces soil erosion. Small mammal burrows that are abandoned also become new homes for countless other species of organisms including salamanders, frogs, and snakes.

Like the warrie cohune palm tree, the spiny pocket mouse bears sharp spines used to discourage attacks from predators. The spines do not dislodge like porcupine quills, but instead, remain attached to the mouse. The warrie cohune palm and its nuts are also heavily protected by sharp spines designed to protect the tasty nuts from being consumed before maturity. Interestingly, the spiny pocket mouse feeds on the spiny nuts of the spiny warrie cohune and is considered one of the most important mechanisms for scattering the seeds of the palm to new locations.

## Birds

More than 350 species of birds have been documented in the BNR and upwards of 375-400 species are predicted to occur in the surrounding Maya Mountains. Wild populations of harpy eagles (one of the largest raptors on Earth, next page), still survive in remote stretches of the BNR, feeding on spider and howler monkeys, tamanduas, white-nosed coatis, and opossums snatched from the treetops and ground. Bat fal-

cons that specialize in catching bats in flight are an amazing feat to observe. All three species of toucans known from Belize are found in the Bladen Nature Reserve, including the outlandish-looking collared aracari.

At least eight species of owls reside in the Bladen, including the unusual-looking spectacled owl, three species of eagles, fifteen species of hawks and falcons, four kinds of vultures including the spectacularly-colored king vulture, ten woodpecker species, eight parrot species, and five kinds of kingfishers that patrol and snare fish from the clear waters of the Bladen River. During winter months, a number of neotropical birds, including the Kentucky warbler, American redstart, cerulean warbler, rose-breasted grosbeak, and the wood thrush are common residents in the deep jungles of the Bladen River Basin. The flamboyant-looking royal flycatcher belongs to the family Tyrannidae, the largest family of birds in the New World tropics.

The slate-colored solitaire, a plain bird of color but an extraordinary bird by song, lives at higher elevation forests. Considered by many to be one of the most accomplished songsters in the world, solitaires produce a clear, flute-like melody that is harmonious to the ears as it carries through the cloud-shrouded peaks of the Maya Mountains.

Other species of particular interest include four species of trogons, the iridescent blue-crowned motmot, the lovely cotinga, several species of euphonias, and nearly 20 varieties of hummingbirds, the jewels of the canopy. Hummingbirds are the only birds that can fly backwards and sideways while beating their wings between 40 and 90 times per second, depending on the species.

Known to the ancient Maya as the sun-God, hummingbirds were believed to be the bringers of first light. By nightfall, the small birds enter into a state of torpor, shutting down all body functions to a bare minimum. The next morning, the delicate birds awake from sleep, bringing with them the first rays of sunlight, and all that it has to offer the Earth so goes the belief of the ancient Maya.

Arguably one of the most interesting birds in the Bladen Nature Reserve, in terms of its song, nest-building skills, and appearance, is the Montezuma's oropendola (referred to as yellow-tail in Belize). Just upriver from the BFREE crossing was a rookery of around 100 nests, hanging high above the tranquil Bladen River. The extraordinary song of this outlandish-looking bird consists of eerie vocalization that begins with a few high-pitched thin notes followed by a sound like the crumpling of brittle paper and ends with an explosive, rich hollow gurgling that carries for a considerable distance.

The distinctive song is produced while the bird falls forward, remaining attached to the limb in a flipping motion. Considered a signature species of the rainforest, the bird's nest-building skills are legendary. The four-foot-long nests are constructed from thin strips of palms, vines, lianas, and other forest debris, skillfully woven into safe and sound brooding baskets. Like many of the bat species in the Maya Mountains, the oropendola is a frugivore, eating a variety of wild figs, sapodilla fruits, and mammay apples, therefore considered an important agent of seed dispersal in tropical rainforests.

The great potoo and common paraque are cryptically patterned nocturnal birds in the family known as nightjars. In low lighting, these birds can see twice as well as other birds, including owls. They feed at night using their enormous mouths to catch insects. The paraque has modified whiskers surrounding its bill to detect bugs, which it promptly consumes in flight. On full moon nights, the eerie call of the common and great potoo resonates through the jungle.

The king vulture, with its nearly 7-foot wingspan, is without question the most colorful buzzard of the Americas. Known to the ancient Maya as simply "Ack", legend has it that as long as this magnificent bird soars over the jungle then all is well with nature. King vultures are, in fact, good indicators of healthy eco-systems, and when land is cleared of trees, "Ack" quietly and sadly disappears. It is no wonder that the best populations left in Central America are found in the continuous forests of the Maya Mountains in Belize.

The yellow-headed parrot and the scarlet macaw are two of the most endangered birds of Belize. Yellow-headed parrots (a regional endemic) are a species of the pine savanna but makes regular visits to feed in the Bladen Nature Reserve. The scarlet macaw (arguably one of the most colorful birds on Earth) also uses the reserve extensively and during winter months, up to 75 macaws (one-third of the known population in Belize) have been observed feeding on the fruits of prickly yellow and other tropical tree species in the Bladen. Unfortunately, both birds have suffered irreparable damage, a result of Guatemalan xateros and Belizeans hunting the birds for the pet trade, deforestation, and damming of the Macal River (the primary breeding ground of the scarlet macaw in Belize). Sadly, these two birds are slipping away, and if not earnestly protected immediately, will suffer the fate of the Caribbean monk seal!

**Serpents**

No other group of animals provokes such widespread panic and fear in humans as do snakes. For some, just the word is enough to send them running. The fear of snakes is called ophidiophobia. While the fer-de-lance and coral snake are legendary for their highly toxic venoms, in my experience, neither of these serpents show an aggressive behavior toward non-prey animals such as humans; most remain tranquil or simply move away. Some of the most specialized snakes on Earth are found here, including the *Sibons* or snail-suckers. *Sibons* are adapted for eating snails, with the lower right mandible equipped with a larger number of teeth than the lower-left mandible. This evolutionary marvel is a result of the dextral land snails (right-sided aperture) that predominate the region. Two of the most common *Sibon* species in the Bladen are *S. nebulata* and *S. sartorii*. To extract the snail flesh, *Sibons* infuse tranquilizing saliva into the snail's tissue to relax its body, which is then extracted from the shell and eaten, leaving the shell undamaged.

There are a number of pencil-thin vine and tree snakes, all of which are equipped with rear fangs and mild venoms for subduing their prey of lizards and treefrogs. Another interesting snake is the eyelash viper, an arboreal snake named for the pronounced tubercles over the eyes which give an eyelash appearance. Eyelash vipers have a prehensile tail used in climbing but more importantly, used as an anchor during a strike and consumption of prey. They eat a variety of small animals including birds, rodents, and even bats. They are considered uncommon snakes of the BNR and occur on both the lower limestone hills and the much higher volcanic mountains of the reserve.

Numerous native snakes are in danger of disappearing forever and while many would lose little sleep over the loss of serpents, ecosystems would suffer immensely. According to the Royal Society, snake populations worldwide are declining at unprecedented rates from gratuitous killing and habitat loss. Another concerning threat to snakes is a fungus called *Ophidiomyces ophiodiicola.* The fungus can cause severe skin lesions and mortality rates can be as high as 90 percent.

The tongue of a snake picks up scent particles which are transferred to the Jacobson's-organ located on the roof of the snake's mouth where it is analyzed for chemical content. The nasal passageway of a snake is also used to sample volatile odors. While the snake's sense of smell is excellent, its sense of hearing is a different story. Snakes are without eardrums but they still possess inner ears. Research has shown that snakes respond to vibrational stimuli that occur on the ground, in air, or even through water. So the popular notion that snakes are deaf is not exactly true, although their hearing is restricted to a limited range of sound frequencies from approximately 50-1000 Hz. While snakes might perceive a stampede of peccaries coming, the high-pitched screams of a panic-stricken human will likely go unnoticed.

When snakes are ready to shed, fluid from the lymphatic system spreads under the skin, separating the old skin from the new, giving the snake a gray or bluish cast. There is a local myth that snakes go blind during the dog days of August. But snakes can shed at any time during spring, summer, or fall and healthy snakes shed more often than poorly fed ones. Furthermore, snakes during this time may be more irritable, a result of being nearly blind by the opaque condition of their eyes.

## Snake Venom

The term poisonous snake is actually misleading. Technically, the substance these snakes possess is modified saliva called venom used to inject, kill, and digest the snake's prey and, to a lesser degree, for self-defense. Venoms are of two basic types: neurotoxin or hemotoxin, or sometimes a mixture of the two. Neurotoxins attack and disable the central nervous system, causing the prey's muscles like the heart and diaphragm to stop working. Hemotoxins target the circulatory system by affecting clotting compounds resulting in uncontrolled bleeding and by destroying subcutaneous and muscle tissue on a massive scale. Unlike venom, true poison can kill by simple ingestion. For example, strychnine is highly toxic to humans and if taken internally can kill within hours. In contrast, venom is generally not dangerous when ingested, having little or no ill effects as long as there are no open sores or ulcers along its path to the stomach where it is neutralized by our stomach's soup of digestive juices.

Another little-known fact is that 40 to 50 percent of venomous snake bites are actually dry bites, which means no venom is injected during a strike (bite). Snakes can and do control the venom flow. This is because making new venom is costly for snakes, requiring multiple meals. This may have started the myth that certain forest plants like monkey-ladder can heal snake bites but there is no scientific evidence supporting these claims. When promptly treated with anti-venom (the only proven treatment for a venomous snake bite), few people die. In Belize, more people depart this life from kissing bugs, lightning strikes, rabid dogs, car accidents, and drowning than from snake bites.

In general, snakes kill food in several ways: by grabbing their prey and consuming it alive like the coffee snake, seizing and squeezing the prey until death like the boa constrictor, or by hypodermic injection of a lethal dose of saliva like the coral snake.

The fer-de-lance is considered one of the more evolved serpents in Belize, having highly efficient methods for killing and hunting prey. This includes copious venom glands connected to enlarged hollow fangs that act like a hypodermic needle and the remarkably evolved pit. The pit is a heat-sensing organ that produces a thermal image, which enables the snake to track a freshly bitten, warm-blooded animal through dense jungle in total darkness. Many snakes in Belize are opisthoglyphous (rear-fanged and only slightly venomous) and bites from these animals may produce local pain and other symptoms of moderate envenomation but are not fatal to humans. Most snakes in Belize are without strong reactive saliva, and therefore, considered harmless to us.

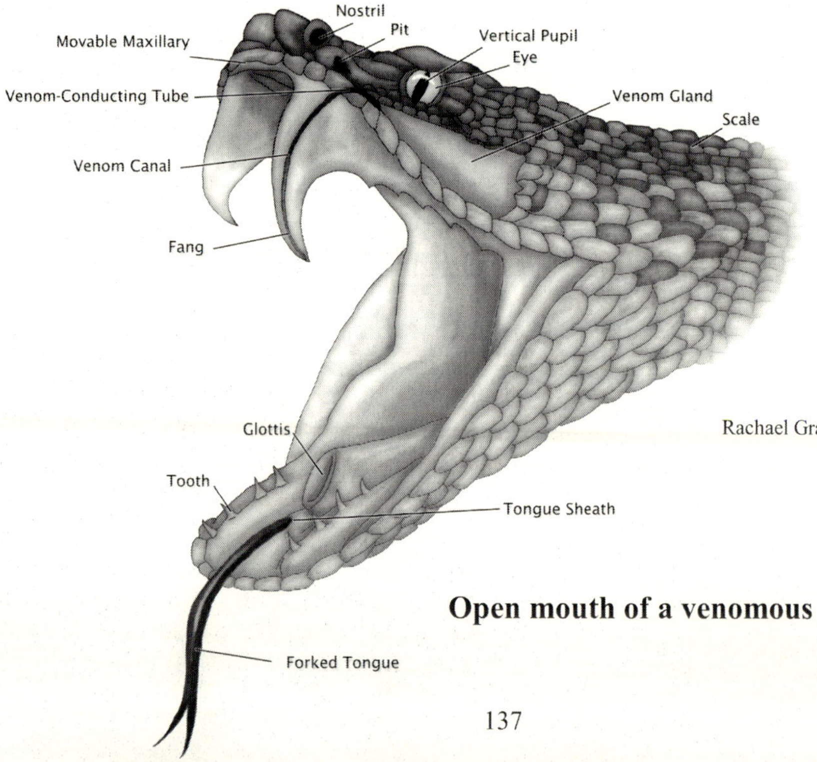


Rachael Grabowski

**Open mouth of a venomous snake**

### Venom Research

Much research has been devoted to snake venom with great promise for developing lifesaving drugs. Toxic compounds such as anticoagulants, antiplatelets, and anti-hypertensive agents found in venom may yield new drugs to treat everything from crippling strokes to the prevention of cardiovascular and cerebrovascular disease. Neurotoxins, found in coral snakes, could be used to treat brain injuries, strokes, and Alzheimer's disease while blood-thinning compounds found in the venom of the fer-de-lance could be used to treat heart attacks or blood disorders.

French scientists believe a compound found in copperhead venom might be useful in fighting breast cancer. Cobra venom is being investigated to treat Parkinson's disease. Research has already resulted in several new drugs based on snake venom compounds including Aggrastat, a "super aspirin" that prevents blood clots. Scientists are only now realizing the potential contributions that venomous snakes will make to humanity, still yet another reason to protect all creatures great and small.

## 16 Common Snake Myths

1. **A snake must coil before a strike.** While a snake can bite from any position, a coiled snake has the advantage of being able to strike farther (typically one third its body lengths).
2. **Snakes are slimy.** Snakes actually have very dry skin.
3. **Snakes are poisonous.** Snakes are venomous not poisonous, their venom is actually, a modified saliva.
4. **Rattlesnakes always rattle before they strike.** Not always.
5. **Snakes will grow into separate fully functional individuals when cut into pieces.** This is completely false.
6. **Venomous snake bites are always venomous.** Up to 50% of venomous snake bites to non-prey targets like deer and humans may be dry bites, meaning no venom is injected.
7. **Constrictors asphyxiate their prey.** It turns out that constrictors kill by stopping the prey's blood flow, starving all organs of oxygen, not by suffocation.
8. **Snakes will chase you.** Very rarely. When snakes feel threatened, they employ 3 strategies to stay alive: remain still, flee, or bite—a similar response from most animals including humans. Racers will occasionally chase and bite a perceived threat but most snakes in general bite only as a last resort.
9. **Some snakes can sting with their tail.** No snakes have stinging tails.

10. **Hoop snakes can bite their tail and roll down the hill.** Anyone that believes this stand on you head!
11. **Snakes cannot strike underwater**: Many watersnakes rely on their ability to bite and seize prey underwater.
12. **Jumping vipers can jump 30 feet.** While jumping vipers sometimes move an inch or two during a forceful strike, they are incapable of launching their heavy bodies any farther unless their striking from a high bank or cliff edge and fall.
13. **Snakes hypnotize their prey.** There is no scientific basis for this silly claim.
14. **Moth balls will keep snakes away.** No, but moth balls are toxic to your pets and other wildlife.
15. **The Fer-de-Lance is an aggressive snake.** Not in my 7 years experience of catching and moving these highly venomous snakes from around BFREE.
16. **Snakes are evil.** Many believe snakes are inherently evil, an idea found in no less than 44 Bible verses. Genesis 3:14 states "and the Lord God said to the serpent, Because you have done this, cursed are you more than all cattle, and more than every beast of the field, on your belly you will go, and dust you will eat all the days of your life". Not one snake of the more than 3400 species worldwide eats dust, except inadvertently, as do most other creatures including humans. Snakes, like legless lizards, crawl on their bellies because an appendage-free design was more functional, especially in round holes where rodents live. Clearly, the folks who wrote the Bible had no real love for serpents and little understanding of the irreplaceable importance that snakes have in functioning ecosystems. Snakes are no more evil than puppies.

## Anurans (Frogs and Toads)

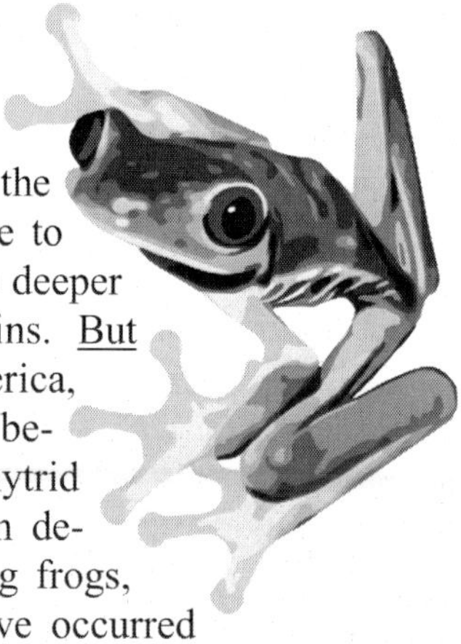

Nearly 40 species of anurans are reported in Belize, a large number of these living in the Bladen Nature Reserve. New species continue to be added to a growing list as inventories push deeper into the largely inaccessible Maya Mountains. But frogs and toads are in trouble. In Central America, anuran declines are now linked to a fungus belonging to the family of saprobes known as chytrid disease. In the past three decades, population declines and extinctions of amphibians including frogs, toads, salamanders, newts, and caecilians have occurred worldwide and amphibians appear to be suffering greater losses than any other group of organisms. Many scientists believe amphibians serve as warning signs and declines in amphibian populations are early symptoms of an ailing planet.

The infamous marine or cane toad, *Bufo marinus,* is the largest anuran found in the BNR. The toad is well-endowed with hefty parotoid glands that contain copious amounts of whitish poison, usually fatal to anything that attempts to include it as a meal. Marine toads are native to Belize, so animals here have learned not to eat this generously proportioned and highly toxic toad. Unfortunately, marine toads have been introduced to countries like Australia for the purpose of controlling cane beetle pests where they have become the dominating nuisance, wreaking havoc on native wildlife populations. Ingested in small quantities, marine toad poison produces hallucinating effects and was used by the ancient Maya in ceremonies and spiritual rituals. These large toads breed and lay eggs in pools found in medium to large river systems throughout the Bladen Nature Reserve and Belize. They can be found in shallow burrows along steep banks of creeks and rivers year-round.

Marvels of evolution, glass frogs are transparent, coin-size anurans with more than 150 described species. Found only in Central and South America these tiny jewels have see-through bellies that display their vital organs such as the heart and intestinal tract. Even more amazing is that most species have green bones.

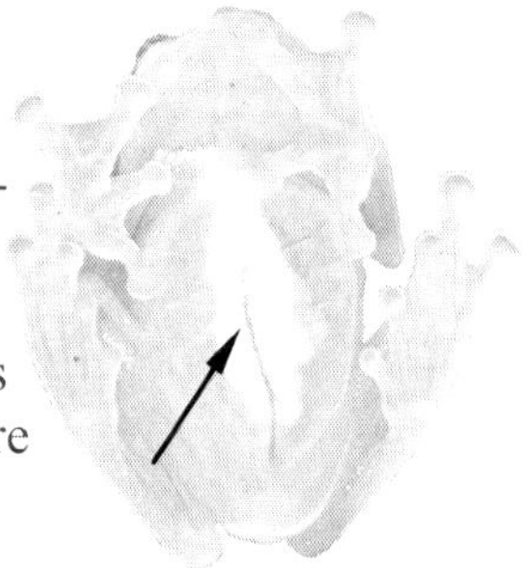

Frogs in the family Hylidae are a colorful group of anurans. Most species are arboreal and are generally referred to as treefrogs. It is a large and heterogeneous family occurring throughout much of North America, Central, and South America. Expanded disks at the tips of the digits (toes) are the family's signature feature. These modified toes allow the frogs to climb vertical surfaces of vines, trees, and even glass windows.

The pepper treefrog, *Trachycephalus venulosus,* is a large tree-dwelling anuran adapted to live in a variety of dissimilar environments including savannas, evergreen jungles, and dry forests from sea level to over 6,000 feet. Most importantly, these frogs are decidedly toxic and therefore, should never be handled without the use of plastic gloves. If disturbed the frog can manufacture copious amounts of a latex-like poison used to protect against predators. There is some evidence that pepper treefrogs are expanding their range in Belize. This is not necessarily good news, suggesting perhaps a bigger, more concerning problem. Research has shown that when plants or animals expand to new locations outside their normal range in relatively short periods of time, it is because other native species are disappearing due to disease or other factors such as habitat loss, deforestation, climate change, or any number of other unanticipated and poorly studied events. Just as disappearing wildlife is an indicator of a planet in trouble, increasing numbers of certain native species can also imply an ecosystem in distress.

### Salamanders

Often mistaken for lizards, salamanders are amphibians that have smooth, moist skin while lizards and skinks have dry scaly skin and sharp claws at the ends of each toe. In Central America, the feet of *Bolitoglossa* species are smooth and paddle-like which aid in climbing vertical surfaces such as trees and vines. Some species like Odonnell's salamander, *Bolitoglossa odonnelli* from Guatemala have a fifth arm; a dexterous, prehensile tail that enables them to hang like a monkey.

Lizards and skinks are generally found in drier habitats and are often observed basking in the sun while salamanders hide in the shadows, under logs, and in other moist places during the heat of the day. Mostly active at night and during the rainy season, tropical salamanders are rarely seen, remaining well-concealed in the canopies of trees and other vegetation. Some species lay their eggs in bromeliads and other aerial habitats but little is known of their entire life cycles. As in most species of salamanders, the color patterns of these arboreal amphibians vary greatly.

Salamanders are poorly known in the Maya Mountains and new taxa await discovery, especially in the higher elevation forests. A recent 2012 expedition into the Bladen Nature Reserve found an intriguing *Bolitoglossa* species, dubbed the painted salamander. Discovered in wet montane oak forests at around 1000 meters, the salamander is currently being studied and likely represents an animal unknown to science. Only two specimens were revealed and both were found in curled up (dead) palm leaves on the forest floor. This may also be the site where eggs are deposited, but this remains to be investigated. Discoveries in the region of the Maya Mountains continue to amaze and excite biologists from across the globe.

### Lizards

Lizards, geckos, and skinks of every shape and size abound in the Bladen Nature Reserve from the nearly seven-foot-long green iguana to the strangely-crested helmeted basilisk. The legendary Jesus lizard of Central America is common in the BNR. This extraordinary lizard is capable of running across the water on its hind legs, an achievement that is hard to accept as fact but as they say, "Seeing is believing!" Back feet are designed to trap small pockets of air, keeping the lizard afloat during its race across the water to flee from predators.

With eyes built to penetrate the inky-black night, magnet-like expanding toes for climbing vertically or upside down, and a detachable tail, the turnip-tail gecko, *Thecadactylus rapicauda* is designed for nocturnal hunting and escaping predators. If caught, the gecko may lose its tail at the point of constriction. If it escapes, it simply grows a new one, although not as handsome as the original model. The predator may miss its quarry but gets a fat juicy tail for its trouble. Interestingly, geckos make chirping-like sounds, whereas lizards and skinks are generally voiceless.

Skinks and lizards appear similar yet differ in several important ways. Skinks have proportionally smaller posterior limbs and in some species, no external limbs at all. Conversely, lizards have long posterior limbs, some species massively disproportionate to their bodies. Skinks have smooth, polished cycloid-type scales whereas the scales of lizards including geckos are generally rough. All species of skinks found in the Bladen Nature Reserve are capable of caudal autonomy and tail regeneration. If toes or tail segments are lost from predator attacks or some other injury, they can re-grow these parts, although not as fully formed as the original ones.

## Crocodiles: Evolutionary Precision

Morelet's crocodiles, *Crocodylus moreleti* (the only species found in the BNR and Maya Mountains) are among the most highly engineered organisms on Earth, having changed little since the days of the dinosaurs. They are long-lived animals that build nests and nurture young, a behavior shared by few reptiles. This brings up an interesting point. Crocodilians are no longer grouped with reptiles but instead, considered more closely related to birds who also build nests and protect their young. Furthermore, crocodilians have four-chambered hearts like birds, whereas reptiles have only three. Although animal shapes are largely a result of lineages, environmental forces also influence an animal's exterior. Dolphins and whales have a fish-like appearance but are in fact, mammals.

Even though Morelet's crocodiles are one of the largest predators found in the BNR, their prey consists mostly of small quarries. Research in Belize has shown that apple snails constitute a large part of the Morelet's crocodile's diet and that consumption of snails increases with an increase in body size. In fact, in the largest crocodiles studied, 70% of all food ingested were golf ball-sized apple snails. The species is considered highly endangered in many areas of Central America and in much of its former range Morelet's crocodiles have been extirpated, a result of overhunting. The great animals are killed and skinned for the hides (primarily used for leather to make handbags, wallets, and shoes), bodies discarded like rubbish—an indifferent conclusion to a remarkable animal that has stood the test of time. Other environmental pressures include the loss of aquatic gastropods due to water pollution, dredging, and habitat degradation, all of which threaten the future of Morelet's crocodiles in Central America.

## Turtles

Seven species of turtles are known in the Bladen River, its tributaries, and the numerous backwater lagoons. The curiously shaped *Staurotypus triporcatus* or giant musk turtle is an engineering marvel. The species is normally found in deep river pools where it feeds on aquatic invertebrates especially mollusks, large amounts of leaves, stems, seeds, and fallen fruit. The three distinct ridges on the carapace are believed to reinforce the shell's strength against attacks from the crushing jaws of jaguars which are known to hunt and feed on turtles. Giant musk turtles can be particularly aggressive and may bite viciously when handled. They are reported to have driven people overboard when caught and hauled into small dories or dugout canoes.

The Central American snapping turtle is the largest freshwater turtle in Belize and is also well known for its defensive behavior, biting brutally if provoked. Diet is much the same as the giant musk turtle. The smaller *Kinosternon leucostomum* or white-lipped mud turtle is far less intimidating mostly found in muddy backwater sloughs and streams, especially in forested areas. Diet includes a variety of invertebrates and plant material but it is also reported to feed on dead fish and other carrion.

## Invertebrates: Earth's Greatest Animal Diversity

The greatest animal diversity on Earth belongs not to mammals nor birds nor even reptiles. That distinction goes to the animals we seldom give a second thought to unless they are dining on our blood or eating our garden. They represent a staggering 95% of the total animal species recorded. For their size, they are faster, stronger, can jump farther, and occasionally, live longer than mammals and birds. Their life cycles are generally more complex and interesting than that of higher-order animals. These amazing creatures are without a backbone and include the well-known arthropods referred to as insects (having six legs), arachnids (spiders and scorpions, having 8 legs), the not-so-familiar myriapods (millipedes and centipedes having multiple legs), and the non-arthropod invertebrates (mollusks). See the pie-chart below.

**Animal Biodiversity Chart**

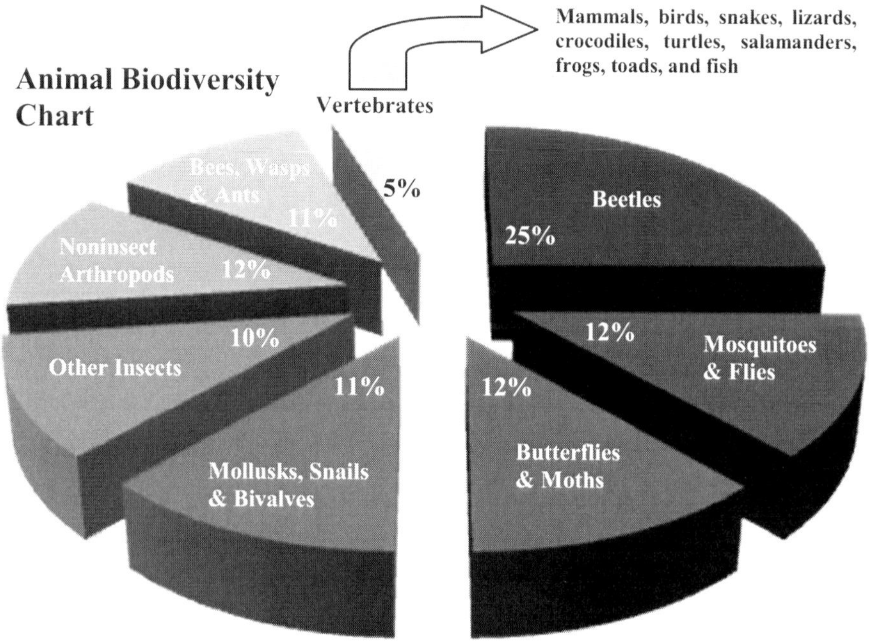


Clearly, along with plants, invertebrates are the driving engines of life. Without the significant and largely uncalculated services provided by these organisms, natural life cycles would slow down and cease to function properly. Many animals, such as birds and bats, insects, and arachnids are the power bars of the food web, especially important to neotropical birds before their migration northward. Insects such as bees are vital pollinators for numerous species of plants and trees.

The most successful animal life forms on the planet today are clearly beetles, comprising 25% of all animal diversity. But in terms of absolute biomass, ants alone make up roughly one-third of the entire animal weight in a jungle. Another remarkable invertebrate is the land snail. They are hermaphrodites (an individual having both sexes), carry love darts, and in rare cases, capable of self-fertilization (producing fertile eggs and young). In times of drought and famine, land snails can go into a state of aestivation, slowing down all body functions for months or even years only to become active when conditions become favorable. Is it any wonder, they've been on Earth much longer and have survived a more precarious, shifting planet? With such diversity and adaptability, some say arthropods will inherit the Earth, while entomologists emphatically state that they already own it. Just look at the pie chart!

The number of invertebrate species remains unknown, likely 10 times the recorded fauna. Since Belize lies in northern Central America, part of the greatest biological isthmus or land bridge on Earth, as such shelters a remarkable quantity of invertebrates, the number of which is yet to be determined. Arthropods are curious creatures but largely the unappreciated underdogs of the natural world. This is because they are scarcely mentioned in other books and for many people, arthropods remain mysterious and threatening. While animals like jaguars and harpy eagles are highly protected by national and international laws, arthropods have little or no official shelter of protection, even though they are clearly the most important players in terms of healthy, functioning ecosystems. Without question, these amazing creatures are in dire need of further study and admiration!

**Beetles**

There are roughly 350,000 currently named species of beetles living in the world today, more than any other group of animals. The total number of neotropical species is unknown but predicted to be astronomical, the majority of which still remain well hidden in the rainforests of Central and South America. Beetles come in every size, shape, and color imaginable, (many beyond belief) and contain some of the largest and smallest living terrestrial invertebrates in the world. For example, the most diminutive beetle on Earth is the feather-winged beetle, *Nanosella fungi* which live in fungus fruiting bodies of mushrooms. At only 0.4 millimeters, it is a speck of dust next to the forest giant Hercules beetle, *Dynastes hercules*, a scarab reaching more than 16 centimeters long (nearly 7 inches) including its outsized horn.

Despite their large diversity, adult beetles of different species are quite similar in that they all contain a hard exoskeleton, well-developed mouthparts, and forewings modified into horny or leathery shells referred to as elytra. Many species of beetles in the genus *Chrysina* or jewel scarabs are extraordinarily colorful. These highly polished beetles wear a multitude of colors including white gold, silver, burnished copper, and shining steel. Others possess porcelain-like surfaces exhibiting dazzling reflections. These living jewels are mostly found in montane oak and pine forests and are sometimes seen around artificial lighting sources.

The larvae of beetles are rarely seen and appear entirely different than adults in terms of shape, color, and food habits. Many are predacious, while others live in rotting wood in advanced stages of decay feeding on wood cellulose. In Peru, beetle larvae (some the size of a person's closed fist) are highly prized as a protein-rich food source.

## Scarabs

Celebrated members of this large and interesting family of beetles include the sacred scarabs of ancient Egypt. In the Americas, scarabs are a diverse group and most of the species belong to five subfamilies: the dung scarabs (1100 species); horned scarabs (620 species); June beetles or chafers (1500 species); flower scarabs (200 species) and shiny or precious metal scarabs (1200 species). Most scarabs are small to medium-sized beetles but some are among the largest insects on Earth. The best characteristic feature of the group as a whole is the club of the otherwise straight antennae, composed of three to nine flattened plates.

## Bugs That Roll Shit for a Living

Dung beetles are one of the more interesting groups of scarabs found in the BNR. Feeding on animal excrement, dung beetles are the cleanup crews of fecal waste. Adults, attracted to the smell of fresh dung, feed directly on or remove portions in which an egg is deposited. Some dung beetles will first form a round ball of excrement which is then pushed backward by the beetle's enlarged hind legs to a particular site where it is buried with an egg. When the larva hatches, a mouthwatering meal is but a bite away. Research has shown that mammal diversity and forest health can be estimated by the number and diversity of dung beetles present.

Illustration by Chris Shields

## The World's Strongest Animals

Among the largest insects in the world, beetles in the scarab subfamily Dynastinae There are roughly 800 species of *Dynastines* recorded from the New World, most occurring in tropical rainforests. The subfamily contains many species displaying remarkable sexual dimorphism, the males generally having horns and protuberances on the head and thorax.

These features are used in battle to defend small individual reproductive and feeding territories. The horn of this large beetle is believed by some indigenous people to act as an aphrodisiac, but there is no scientific basis for this claim. Some also believe that the beetle is capable of cutting off the limbs of trees using its enlarged horns as a saw, a fictitious but interesting anecdote. The massive larvae feed on rotting wood and pith of various trees, including palms, and may act as significant nutrient recyclers. Several species may be indicative of old-growth forests. Research has also shown that scarab beetle assemblages can help predict areas of endemism.

For their size, members of this family are the strongest animals on the planet, a number of species capable of lifting 850 times their body weight, the equivalent of a human lifting over 60 tons. Hogue reported that a Hercules beetle he kept in a birdcage would escape at will by simply prying apart the steel bars. A comprehensive biodiversity inventory of the Dynastinae of Belize, including Mexico and Guatemala by Ratcliffe, Cave, and Cano is currently in print.

## Talking Bugs

Passalus beetles, often called bessbugs are highly developed subsocial insects in which the adults and developing larvae live together in small groups. Both stages produce a repertoire of sounds that can be heard by picking the beetle up and holding it close to one's ear. Fourteen different acoustic signals have been recognized, 11 of which are associated with specific behavior, the most reported for any arthropod species. Typically found in rotting wood and under logs, their diet consists of decaying wood fiber and their decomposing feces, yummy! The beetles themselves are inhabited by ectoparasitic and phoretic mites (easily seen with a hand lens) that are typically located on the undersides of the adult beetles.

## Leafy-Bugs: Katydids

These bizarre-looking arthropods belong to a remarkable group of illusory insects referred to as leaf katydids. Their outer wings are precise imitations of green, dying or dead leaves, complete with drip-tips, midribs with side veins and discolored areas resembling insect feeding holes, chewed edges, leaf miner tunnels, and mold spots, all of which completes the deception that they are a leaf, not an edible insect. Because they are arboreal and most active at night, they are difficult to spot and few people ever see these well-camouflaged insects. Little is known of their biology and nearly 100 species are recognized from Central and South America. Furthermore, research has shown that katydids are particularly important food sources to a host of vertebrate species including birds, bats, monkeys, lizards, and young snakes.

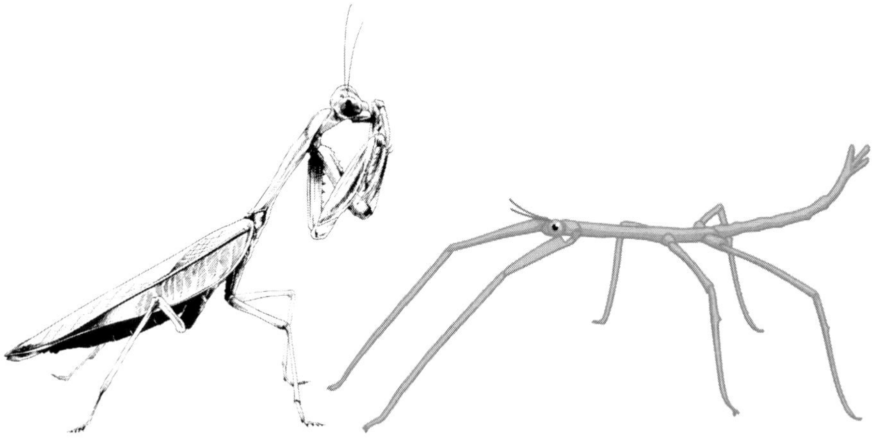

## Insects that Pray and Sticks that Walk

The praying mantis gets its common name from the position in which it holds its greatly enlarged forelegs. But these arms are not used for prayer but prey. They are lethal weapons, armed with spines, with vise-like grips used to grab and hold their quarry during consumption. All mantises are hunters, taking an assortment of insects, and have been known to even catch and eat small vertebrates such as hummingbirds. They are a well-represented family in neotropical regions with more than 300 reported species, and many new taxa await discovery.

While mantids are hunters of flesh, walking sticks are seekers of cellulose. Walking sticks are the twig mimics of the insect world blending superbly with their surroundings, so are seldom seen. This is their first line of defense but a number of species are equipped with a pair of glands at the anterior edge of the prothorax that enables the insect to release defensive secretions (caustic chemicals) that can blister human skin, or have the ability to change color as their surroundings shift. Among the longest insects known, several South American species measure over 30 centimeters (12 inches long). Interestingly, walking sticks can regenerate lost limbs.

## Cockroaches of a Different Nature

In the tropics, native cockroaches (not to be confused with the few exotic pest species that invade our homes) are one of the most diverse and biologically fascinating groups of insects in the world. Cockroaches are a primitive order with ancestry dating back some 250 million years. The basic body design of cockroaches has changed little since then. All cockroaches have a thin layer of grease or a waxy covering on the outside of the cuticle which protects them from desiccation or drying and many species secrete a variety of exocrine substances used in defense. Cockroaches from the family Blattidae have an adhesive amalgam used in defense against would-be predators; anything attempting to include the roach as a meal receives a mouth full of glutinous substance instead of a scrumptious meal.

The alimentary canal of several species of wood-feeding cockroaches contains symbiotic protozoa which, like those of termites, assist in the digestion of wood fiber. Native cockroaches help reduce the accumulation of leaf litter and wood, recycling vital nutrients back into the environment. The giant cockroach, *Blaberus giganteus* is the largest native species in Belize, some specimens reaching 8-10 centimeters in length. It has long spiny legs used for climbing and in defense. The large head shield, seen in most cockroaches, is used to protect it from predator attacks. The giant cockroach can make hissing noises and feeds on detritus, bat guano, decomposing vegetation, and rotting wood. It is a common roach in tropical jungles and is also a frequent resident of buildings.

## The Dragon-headed Bug

The dragon-headed or peanut-headed bug, *Phenax variegata,* is arguably one of the strangest-looking animals in the Bladen Nature Reserve. Using mimicry, this large planthopper resembles a tree-dwelling lizard. Its large sized (nearly 5 inches), false head, counterfeit eyes, elongate form, and mottled green and white wings all assist in the deception. Lizards will falsely recognize the dragon -headed bug as one of its own kind and usually avoid eating it. Much folklore surrounds this group of sizeable and oddly-shaped insects, including the belief that it is venomous and consequently, deadly. There's a widespread myth that anyone bitten by this creature must have sexual relations within twenty-four hours or suffer a horrible death—no doubt a notion initiated by a male *Homo siphon* having difficulty with the opposite gender. In truth, dragon-headed bugs pose no threat to humans and do not bite. The bugs do, however, feed on trees and shrubs that contain bitter and resinous sap. It is thought that harmful substances from the sap are sequestered by the bugs and used as a secondary defense behind their lizard-like mask. The bug has one more trick up its sleeve, (actually below its wings). If all else fails, the dragon-headed bug protects itself by flashing its large but phony eyespots, located on the corners of the hind wings.

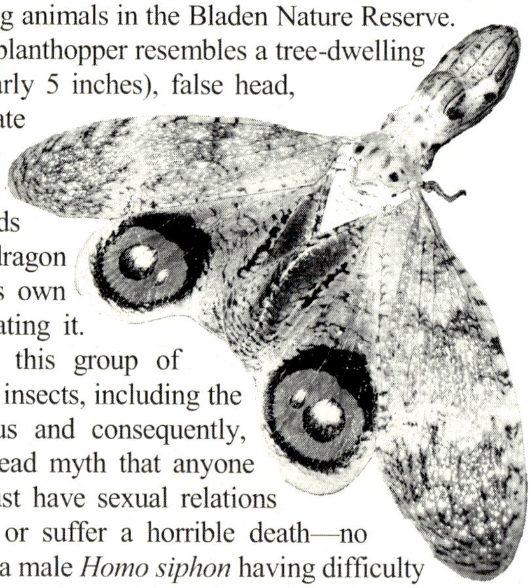

## Grasshoppers and Crickets

Grasshoppers are so familiar that they need little description, differing from crickets and katydids primarily by having shorter, thicker antennae. Most grasshoppers are capable of making loud sounds by frictional contact between various parts of the body such as the hind legs and wings. Grasshoppers in the family Romaleidae are found only in the New World and are the largest grasshoppers on Earth (body lengths up to 105 millimeters in Belize). Forty-eight genera are reported with around 100 species. Many are flightless, having reduced wings, and are colorfully marked, warning of repugnant secretions used in defense.

Crickets are best known for their musical talents; their chirps and trills more melodious than the rasping clatter of katydids and the buzz-like calls of grasshoppers. One of the more interesting crickets of the Maya Mountains is the mole cricket found living along the sandy banks of the Bladen River. This cricket nests in subterranean galleries it constructs using massive digging chisels and scrapers. Like most species of crickets, male mole crickets use songs to lure mates from burrows and other hiding places.

## Cicadas

Cicadas are among the most famous insects in the world, a result of their capacity to produce ear-piercing sounds. Often wrongly referred to as locusts (which are actually grasshoppers), cicadas are common inhabitants of forested regions. The male's seductive buzzing attracts females for mating after which the female inserts masses of eggs into twigs with its ovipositor. Hatching nymphs drop to the ground and burrow deep into the soil where they remain for two or more years, feeding on the sap of tree roots. At maturity, nymphs emerge from the ground sometimes in mass numbers, crawling up trees and shedding their nymphal exoskeleton to become adults where their life cycle starts all over again. The surplus cicadas become an important source of food for numerous animals such as lizards and birds. Over 800 species are reported from Latin America alone.

## Charmers of the Insect World

Lepidoptera (butterflies and moths) are a sizeable group of arthropods in the tropics but have remained largely unstudied in Belize. These are the charmers of the insect world and like birds, many people spend a great deal of time and money searching for these winged jewels. The Bladen Nature Reserve is home to a large quantity and the number of species is yet to be determined but is predicted to be substantial. Many newly discovered areas in the reserve such as Oak Ridge, the Lost Valley (found while on 2011 and 2012 expeditions), and areas along the largely inaccessible higher elevation Maya Mountains remain entirely unsampled for this diverse order of insects. Lower elevation regions of the reserve (examples including Snake Creek and Pass Valley) remain unknown as well. No doubt numerous species new to science are awaiting discovery.

Considerable wing shape and color diversity exist in Belize species, each wing shape giving a particular species of butterfly different flight abilities. The unique wing patterns and colors also aid in crypsis and mate recognition. Butterflies consume a variety of foods including flower nectar but also a host of putrid and often disgusting (to humans) fodder. On the menu is rotting carrion, fresh scat, and an all-time favorite, puddles of fresh urine. The more nauseating the food, the greater number of butterflies are attracted to it. So next time you are in the jungle looking for those winged jewels, try following your nose.

Many species of Lepidoptera have false eyespots located on the underside such as the owl butterfly. Theories about the role of these eyespots have been debated since the 19th century with several hypotheses emerging. Early conjecture included predator-eye mimicry (such as owls), but recent research suggests that conspicuousness, not eye mimicry, may encourage avoidance behavior in predators. Other theories include the idea that the eyespots serve as decoys, diverting attacks away from the vulnerable body parts of the butterfly.

Caterpillars are the larvae of moths and butterflies. They are as varied as the adults both in terms of color and shape. While most caterpillars are cryptically colored, blending well with their surroundings, many species are covered in delicate hairs, sheltered by venomous spines, or are masters of deception. Spiny caterpillars carry powerful and painful stinging spinules that are highly venomous to the touch and few predators will endure the pain for a meal. One of the most remarkable animal deceptions in nature is the viper worm, a caterpillar of the hawkmoth. When threatened, this master of disguise reacts aggressively, turning its phony snake head back and forth toward the aggressor. Perceived as a snake, the potential predator is startled, confused, and in most cases, will look for another meal.

### Mosquito Hawks

Odonata (dragonflies and damselflies) are a large and diverse order of predacious insects frequently seen flying near or over bodies of water. Often referred to as mosquito-hawks, these colorful arthropods are outfitted with large compound eyes, giving them excellent vision, and have two distinct life stages. The first stage is in the form of an aquatic nymph where it hunts invertebrates, small fish, and tadpoles. As adults, dragonflies live in terrestrial environments, next to the water where they hunt pesky doctor-flies and adult mosquitoes. While sitting on vegetation, dragonflies can be distinguished from damselflies by the position in which they hold their wings. Dragonfly wings remain flat while damselfly wings are usually folded together. Like many other groups of animals, little or no work has been done on Odonata species in the Bladen Nature Reserve and surrounding Maya Mountains and many new species are expected to be found.

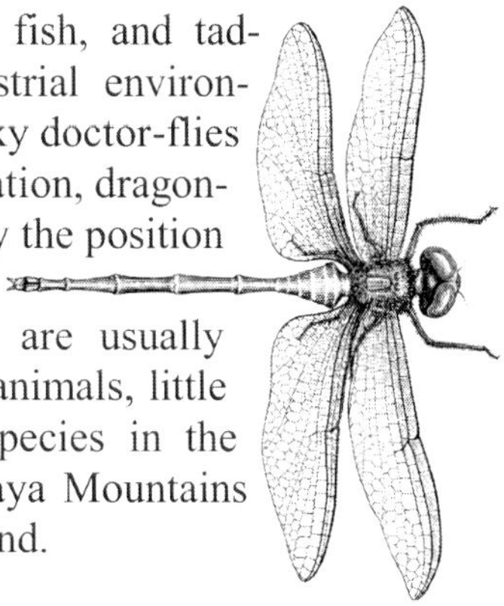

The helicopter damselfly, *Magaloprepus coerulatus,* is the largest in the family Zygopterona in the Americas with a wingspan of more than 160 mm. An inhabitant of mature forests, eggs of this large damselfly are deposited in bromeliad tanks and tree crotches that hold water. Nymphs feed on mosquito larvae and probably the tadpoles of frogs that lay eggs in these container habitats. As it hovers, the adult damselfly feeds on kleptoparasitic spiders snatched from the webs of orb weavers. Locals believe that these damselflies are the spirits of recently deceased humans, insisting that they not be caught or harmed.

## Hymenoptera: Ants, Bees, and Wasps

Several of the most familiar groups of arthropods in the tropics include the Hymenoptera (ants, bees, and wasps). This large and diverse group of insects are especially important in ecosystems in terms of bug control and plant pollination.

## Paper Wasps

Paper wasps are highly social and skilled builders, turning chewed wood fiber (cellulose) into complex nursing chambers for developing eggs and larvae. The largest, social paper-wasp nest in the world is the size of a 30-gallon drum. Nest building strategies of paper wasps are of two basic types. The more simple design has a single exposed comb secured to the underside of a leaf. In the second, more complicated nest, the comb is encased in a protective cover of paper or carton, a wood-based material used by nearly all social wasps.

Drumming wasps, when disturbed, produce rhythmic drumming sounds by vibrating the nest's inner walls. This can be heard several meters away. If this fails to drive off the intruder, then hundreds of individuals emerge onto the outer surface of the nest and continue to make the thumping (warning) sounds.

These wasps are formidable defenders of their hives and have a potent sting. Like honey bees, however, this species of wasp possesses a barbed stinger that is lost during a sting, resulting in death to the wasp. This consequence causes the wasp to be less prone to attack immediately, using instead the drumming reverberation as the first line of defense.

**Mud Daubers**

The mud-dauber is a species of spider-hunting wasp that builds a nest of clay. Once the nest is completed, the wasp catches a spider, paralyzes it with a sting, and carries it back to the nest (still alive). There, the spider is placed inside the nest with one mud-dauber egg attached and then the chamber is sealed. When the egg hatches, the young wasp lives off the living tissue of the paralyzed and unlucky spider.

**Parasol Wasps**

The Parasol wasp (*Apoica* species) is of particular interest due to its nocturnal behavior. During the day the wasp hangs on the nest, looking like an umbrella, hence its common name. Parasol wasps have powerful and painful stings used to defend hives and immobilize prey. This genus of wasps is poorly known, due to its nocturnal activities, and contains only six recognized species, two species found in the Bladen Nature Reserve. They are a species better left alone.

**Tarantula Hawk**

The tarantula hawk is a sizeable (5 to 8 centimeters) hunting wasp. Prey consists entirely of live spiders, including tarantulas, which are hunted down, stung, and paralyzed but not killed. The wasp then drags the tarantula to a burrow, (often the spider's own), lays a single egg on the spider's body, and seals the burrow. The larva feeds externally on the interred carcass of the spider until adulthood. Apparently, *Pepsis* wasps are specific in terms of prey species and proficient at discriminating different spider species by odor clues picked up as the wasp taps the spider with its modified antennae. Members of this genus include several hundred species, all restricted to the New World tropics.

**Orchid and Stingless Bees**

Orchid Bees occur exclusively in South and Central America including Belize. Most species are characterized by brilliant metallic coloration, primarily green, gold, and blue. Male orchid bees have uniquely modified legs which are used to collect and store different volatile compounds from wild orchid plants throughout their lives, largely from orchids in the subtribe Stanhopeinae.

Stingless Bees, or as I like to call them "sticky bees" because they stick to you when encountered, do not sting but are not defenseless and will protect their hives aggressively. Intruders are attacked in large numbers and bees will bite with their large mandibles, crawl into nostrils, hair, ears, and eyes of the offender with several species employing a caustic fluid from glands located at the base of their jaws. We have experienced these creatures hanging on the outdoor shower head, lapping up water. Before you take a shower, you always have to run the water a bit to avoid being showered with bees.

## Ants

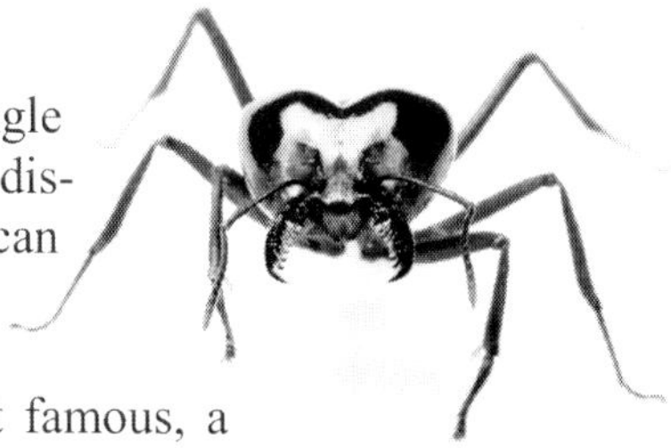

The unparalleled sovereigns of the jungle are not jaguars nor harpy eagles. That distinction goes to the ants. A single tree can support more than 50 species and numbers in some colonies are staggering. The leaf-cutter ants are among the most famous, a result of their legendary ability to farm pure strains of a unique fungus, which is cultured on leaf cuttings and other parts of plants carried to underground nest chambers. The tips of the hyphae (part of the fungi) produce round swellings (gongylidia) that are detached and consumed by the workers or fed to the ant larvae and queen. A single leafcutter ant mound can cover more than 50 square meters, towering a meter high, decades-old and house a staggering eight million individual ants. The enlarged galleries are specially designed to grow the fungi; the ant's only food source. Eggs and larvae are simply embedded throughout the fungi where they have easy access to a constant supply of nourishment. Dark-colored individuals with wings are male ants that perform no labor inside the colony but instead wait to disperse the colonies' genes during periodic mating flights with new queens outside the nest.

## The Velvet Ant (An Imposter)

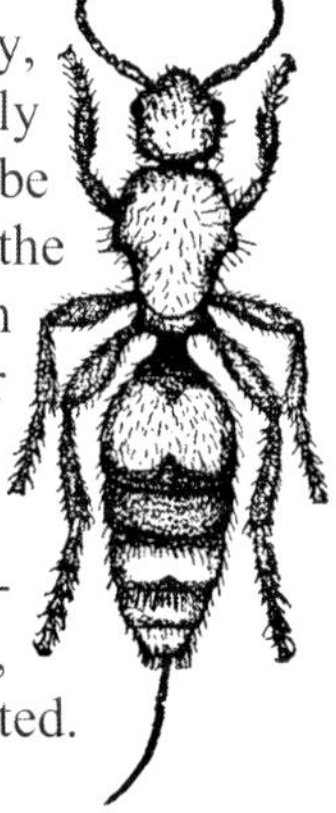

The velvet ant is not an really an ant but a species of solitary, flightless wasp in the family Mutillidae. Containing nearly 1100 recognized species, there are many more expected to be discovered as inventories continue in remote regions of the tropics including Belize. Female velvet ants are often seen walking on the ground and are usually dark velvety blue or black with brightly contrasting red, yellow, or white spots. Many neotropical species are large (up to 20 mm) and lack wings. The excessively long and painful stingers of the females are used to pierce the nest cells of their hosts (bees, wasps, beetles, and flies) where venom and an egg is deposited.

There is some interesting folklore surrounding these remarkable wasps. In Brazil, velvet ants are used in love magic rituals. Desiring the attentions of female companionship, a man obtains three of the women's hairs, coats them with a sugary syrup, and places them in a box with a velvet ant. As the wasp feeds on the hairs, the woman gradually falls in love with the suitor. If it was just that easy!!

### Giant Hunting Ants

The infamous giant hunting ants (sometimes referred to as bullet ants) of Central and South America pack a powerful, painful sting that is compared to the feeling of being shot. Although I have never been shot, I have in fact been stung by a bullet ant and can attest to the ferocity of the claim. These large ants (up to one inch) live in small colonies (a hundred or so with a queen) in burrows located at the base of trees. Workers forage alone, hunting the forest floor and trees for arthropods and plant material to feed their larvae. The ants do not demonstrate group behavior, a trait considered primitive, and are generally restricted to the wetter regions of tropical America.

### Carton Ants

The nests of carton ants look remarkably like the nests of nasute termites but are notably smaller and structurally inferior. Carton ant nests are built in trees and vines comprising two or more individual nests (10 to 30 feet apart), all part of the same family. Primarily made of masticated wood mixed with saliva, the nests are relatively fragile. To stabilize the nest's internal structure, carton ants harvest and plant seeds from epiphytes (bromeliads and orchids), inside the nest. The seeds germinate, grow and spread their roots, giving the nest structural stability. While most of the masticated wood is washed away during heavy rains, the core of the nest survives along with the queen and larvae thanks to the well-established root system of these plants. The nest is simply rebuilt. These small ants are highly aggressive and while they lack the stinging apparatus found in most other ant species, they bite and emit foul orders (resembling butyric acid) at any trespasser.

### Nasute Termite Nests

Like many insects, termites have gotten a bad rap and while a few species become problematic in our dwellings, all species are considered important components in natural and functioning ecosystems around the world. For example, research has shown that termites add significantly to soil aeration and fertility by their tunneling action, digestion, and breakdown of cellulose. Termites rely on symbiotic flagellate protozoa and bacteria for digestion of wood fiber and other complex polysaccharides in their diet. The building material of these large aerial structures consists of a paste from chewed wood and fecal cement and is a much tougher material than that of carton ant nests. Termites themselves provide a protein-rich food source for birds, reptiles, amphibians, and the marvelously evolved ant-eaters of Central and South America. The nests of nasute termites also provide nesting and important roosting sites for a host of species such as birds, reptiles, and several species of bats.

### Who Doesn't Love Spiders?

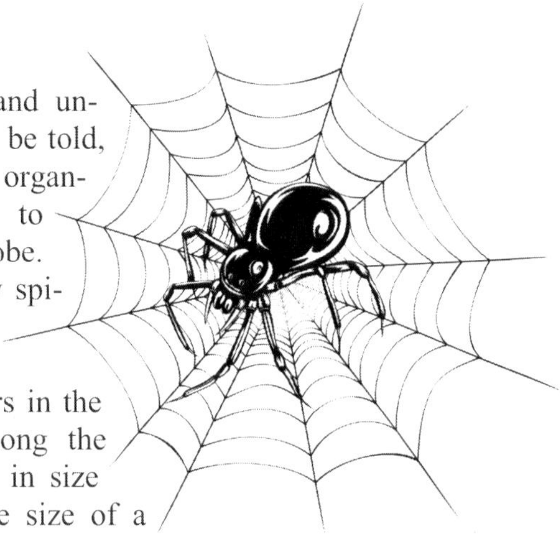

Although spiders provoke fear and unease in many humans, if the truth be told, this large and diverse group of organisms are incredibly important to healthy ecosystems across the globe. Without the services provided by spiders, we would spend untold amounts of money controlling what are considered pests. Spiders in the Bladen Nature Reserve are among the most familiar arachnids, ranging in size from minute jumping spiders the size of a pinhead to tarantulas the size of small dinner plates. From orb weavers to wolf spiders, they are important predators of the forest, keeping many insect populations in check. To appreciate the colossal number of spiders in the rainforest, one need only walk down a jungle path at night with a headlight. While the greenish-blue eye-shine of countless spiders is unsettling to some, I find it to be a captivating experience. Spiders are also known for manufacturing one of the strongest natural fibers known on Earth—silk.

Due to their large size, tarantulas are among the best-known spiders in tropical regions. Although tarantulas are widely looked upon with great fear and believed to be lethal, their venoms are not strong enough to kill humans. Tarantulas will bite in self-defense but are more likely to discourage attacks by flicking the hairs located on the dorsum of the abdomen with one of the hind legs. These hairs are irritating and may lodge in the eyes or sensitive nasal membranes of a would-be predator. By rubbing the surfaces of basal segments of the pedipalps against opposite surfaces of the first legs, some neotropical tarantulas can produce a snakelike hissing sound that can be heard up to 20 feet away. It is unknown how many species exist in the BNR, but there are currently eleven taxa reported from Belize.

## Scorpions and Their Kin

While spiders comprise the bulk of arachnid diversity in the BNR, scorpions are also common, although less speciose. They range in size from the 3 mm long pseudoscorpion that injects venom through its pincers to the six-inch-long forest scorpion that can deliver a knee-buckling sting, although not considered life-threatening. Forest scorpions (the common species found in the BNR) hunt grasshoppers and crickets but also eat vertebrate animals such as lizards. Scorpions grab their prey with vice-like pincers that hold the prey securely and with a sizeable stinger, inject a lethal dose of venom into the victim's body which is then eaten.

In Mexico, members of the genus *Tityus* and *Centruroides* are reported to carry strong neurotoxins, causing circulatory and muscular effects in humans, and are of major medical importance there. Although scorpions found in Belize have painful stings, they are considered relatively harmless to humans.

Interestingly, all true scorpions fluoresce (above image) at night when exposed to certain wavelengths of ultraviolet, the kind produced by a black light. This is a result of fluorescent chemicals present in the cuticle (skin). The function of fluorescence in scorpions is somewhat a mystery and many arachnologists speculate that scorpion fluorescence has no function at all, being simply a random act of evolution. California State University arachnologist Carl Kloock thinks otherwise. His research suggests that scorpions may be using UV as a way to determine whether or not to come to the surface to look for prey, based on the light levels. It turns out scorpions specifically avoid UV light. Moonlight has a component of UV light and research has shown that scorpions are less active during full moon nights and more active during darker nights. According to Kloock, fluorescence may be part of the mechanism by which scorpions respond to moonlight, adjusting their activity level depending on the amount of UV shining on their bodies.

In addition to the familiar scorpions, there are those strangely-shaped vinegarroons (pictured right) found under rocks and Amblypygids or tailless whip-scorpions, common to limestone caves and hollow trees. These oddball arachnids are generally timid creatures; however, when faced with danger, vinegarroons squirt acetic acid compounds from their anus, causing temporary burning and blurred vision. It is best to keep your inquisitive eyes at a respectable distance (more or less around a foot)! Amblypygids are without stingers or chemical defenses so are completely harmless and can be handled without personal harm. These large arachnids are restricted to moist tropical or subtropical forests and there are numerous species. Interestingly, the diets of tailless whip-scorpions include other arachnids such as wolf spiders.

## Millipedes

Millipedes are similar to centipedes but differ in having two pairs of legs from most segments of the body while centipedes have only one pair of legs per segment. The diet of millipedes typically consists of decaying wood and rotting vegetation. Around 750 species have been reported from Mexico and Central America, although the total fauna is thought to be several thousand species, probably less than 20% of which have been discovered, named, and described. Millipedes are generally considered harmless to handle and are protected primarily by a tough exoskeleton. They do, however, produce caustic secretions from a series of lateral pores on most of their segments. The active chemicals of these secretions are varied but include benzoquinones, cresols, and aldehydes that can blister or tan human skin. Polydesmid millipedes, which are common to the Bladen Nature Reserve, secrete hydrogen cyanide. So, if you handle these interesting creatures, be sure to wash off any lingering secretions and remember not to rub your eyes! Like scorpions, several species of millipedes fluoresce under UV light for reasons that remain unknown.

## Centipedes

Centipedes are a different story and should not be handled, having several legs modified into four-jointed, sharply-pointed fangs connected to internal venom glands. Although the bite of the centipede species found in Belize is painful, causing localized inflammation, it is not considered dangerous to humans. When molested, some centipede species will employ strong-smelling chemicals. Unlike millipedes that are generally herbivores, centipedes are nocturnal hunters, eating invertebrates such as earthworms and insects but also bats, mice, lizards, and toads are included in the diet of larger centipede species.

## Better than Fiction—The Velvet Worm

It is often said that fact is better than fiction and this is certainly the case for the velvet worm. These secretive millipede/centipede-like critters are neither, belonging to a primitive family (Onychophora), some 540 million years old. Their velvet-like skin is highly sensitive to the surrounding environment. Although they appear harmless, velvet worms have an ominous nature, possessing highly specialized sliming organs. Two nozzles located below the worm's feelers (antennas) are used to shoot streams of sticky slime to capture prey or ward off attacks from predators. If agitated, the velvet worm will face an adversary and with force, squirt a viscous mucus. Velvet worm prey on termites and other invertebrates. Velvet worms are reported from moist tropical regions of the world and a few cave systems. These nocturnal hunters live around moist leaf litter and rotting wood, but are rarely seen.

## Land Snails

Land snails come in every color, form, and size imaginable; many beyond belief. Take terrestrial gastropods in the genus *Opisthostoma* (illustrated below). These little jewels have taken calcium carbonate sculpturing to a whole new level of architecture, representing some of the most spectacular achievements in convoluted evolution ever seen in a snail. These highly restricted gastropods are found on limestone outcrops and in deep caves of Borneo.

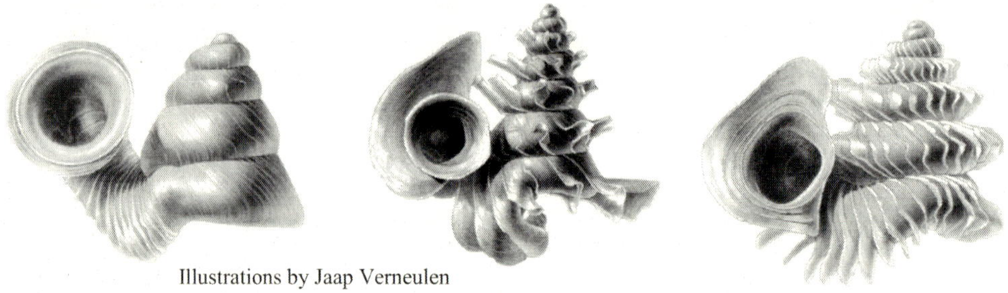

Illustrations by Jaap Verneulen

Snails are part of a large and diverse family of organisms known as Gastropods, all belonging to the enormously diverse Phylum Mollusca. Containing nearly 100,000 described species worldwide, mollusks include aquatic snails (marine and freshwater), conchs, land snails, terrestrial slugs, sea-slugs, sea-hairs, limpets, bivalves (clams, oysters, and mussels), squids, octopuses, and nautiluses. Octopuses are highly-intelligent mollusks having feet divided into a number of prehensile and skillful tentacles capable of twisting the lid off a glass jar to access the food within. Although snails and slugs are slow-moving, squids are among the fastest animals, exceeding underwater speeds of more than 70 mph in short bursts, a result of their jet propulsion. At eighteen meters (nearly 60 feet) and weighing in at around two tons, the giant squids are the largest living invertebrate known to science with eyes the size of basketballs (the largest of any animal). Mollusks are also the longest-lived, multi-celled animals on Earth with one species of bivalve, the ocean quahog, *Arctica islandica* living more than 500 years.

There is scarcely a place on the dry surface of the world, outside the polar regions, where one cannot find at least a few examples of land snails. From hot, nearly waterless deserts to snow-capped mountain-tops, land snails are thriving. Despite such biological and physiological limitations, land snails have developed efficient mechanisms for coping with freezing, starvation, and desiccation. For example, when conditions become increasingly dry, snails cover the aperture of their shells with an epiphragm, a mucous sheet that hardens, sealing in critical moisture, and preventing desiccation. Some snails can remain dormant for years, resuming activity during wet weather. Land snails in Belize are typically nocturnal animals and are largely active in warm (75-85 degrees Fahrenheit), rainy weather.

Even though mollusks rank as one of the most numerous and speciose groups of organisms on Earth, they remain largely unstudied. As a result, little is known of their importance in many ecosystems, and like many invertebrates, suffer from being a conservation "blind-spot." In Central America alone, it has been estimated that as much as 65% of the gastropods await discovery.

As snail research moves forward, however, our understanding of the value of these organisms is increasing. Research has shown that land snails play an important role in micronutrient cycling in terrestrial ecosystems, disperse plant seeds and fungal spores, and can be used to predict vertebrate conservation priorities. Further, live snails and their vacant shells provide a food and calcium carbonate source to many systematic groups. These include but are not limited to ants, firefly larvae, snail-killing flies, *Cychrine* beetles that feed chiefly on land snails, arachnids including harvestmen, carnivorous snails, numerous species of salamanders, turtles and frogs, a variety of small mammals including shrews, mice, and moles, snakes, a variety of passerine birds, thrushes, ruffed grouse and wild turkey, bats, and primates including humans.

While a building body of evidence suggests the importance of mollusks in present-day ecosystems, their historical value is less well-known; especially their contribution to existing plant communities, animals, and, in particular, caves. In the past, the colossal accumulation of deceased mollusks, corals, and other tiny creatures known as Foraminifera that have calcareous skeletons, provided the necessary building material to create limestone where caves are essentially formed. Many species have uniquely evolved in caves that can live nowhere else. Furthermore, a whole host of plant and animal communities are linked or dependent on limestone soils. These ancient shells have also provided the necessary limestone in cement to form the foundations of our cities and homes.

Declining land snail populations can have ripple effects to surrounding ecosystems. The great tit (*Parus major*) in the Netherlands, for example, has declined precipitously with the decline in land snails, as a result of acid rain. A lack of snail shells in the bird's diet causes its eggshells to thin and break, therefore reducing reproductive success rates of the species. In North America, Hames has documented a correlation between a reduced number of wood thrushes (a winter resident of Belize) and acid rain, hypothesizing a connection to reduced land snail populations.

Sensitive to changes in their environment, land snails could provide an early warning to impending habitat deterioration. Research has shown, for example, that when snails feed on contaminated foods such as mushrooms and vegetation, environmental toxins are ingested and sequestered in their tissues, the midgut gland being the main accumulation site of these trace elements. More concerning, however, snails quickly become poisoned when simply raised on soils contaminated with cadmium, raising fear that toxins in polluted soils may be more bio-active than previously believed. The environmental implications to ecosystems and the consequence to snails and higher organisms that feed on contaminated gastropods are valid as well as alarming. Research has shown elevated levels of mercury in the blood of Bicknell's thrush on mountains in New England, USA, and thrushes are reported to eat snails—the possible source of mercury contamination. Native land snails could, therefore, be used to forecast impending problems created by anthropogenic pollutants believed to be accumulating in the jungles of Belize. This hypothesis, however, remains untested in the country.

## Aquatic Snails

Aquatic snails are those gastropods found in Rivers, tributaries, springs, still water lagoons, and vernal pools. Although the diversity of aquatic snails is substantially less than that of land snails, they make up in numbers what they lack in diversity and are considered key organisms in aquatic systems. For example, Cochran concluded that 62% of the diet of multiple fish species (i.e. cichlids, bay snook, and mountain mullet) from the Bladen River came from three aquatic snails, *Pachychilus corvinus*, *Pachychilus largillierti*, and the non-native (Asian) *Melanoides tuberculata.* Aquatic snails also consume large quantities of abscissed leaves (most often *Ficus* species) that fall within their reach. As the snails feed on the leaf tissues of the fig, waste products of the gastropods are released into the water column, spawning a new flush of nutrients. These nutrients become food for a whole host of aquatic microorganisms which are then fed on by fish.

The apple snail is a bulky gastropod found in still waters of lagoons, old oxbows of the Bladen River, and vernal pools. They are among the largest freshwater snails in the world and are an important food source for otters, American and Morelet's crocodiles, fish, freshwater turtles, a variety of egrets, herons, and snail kites. Research has shown that the endangered Morelet's crocodile relies strongly on apple snails as an important and reliable source of food.

River snails were also utilized extensively by early Mayan civilizations (as evident in so many Maya ruins) as a reliable food source. Even today, aquatic snails are used by Belizeans to treat eye problems such as glaucoma. Live snails are gathered from the river, broken in half, (exposing a watery slime) that is applied to the affected eye. The Mayan Quill, *Brachypodella speluncae*, (pictured to the right) is used by the Maya to treat whooping cough. "Tu-tut" (which broadly means snail in Kekchi Maya) are roasted live on a comal (a flat griddle) and mixed with tiny black stones gathered from clay deposits. The mixture is then ground to a powder, mixed with warm water and consumed to cure the uncontrolled whooping cough (pers. comm. Santiago Coc, 2009, San Pedro Columbia, Belize).

Clearly, the loss of aquatic gastropods such as apple snails due to water pollution, dredging, or other forms of habitat degradation would threaten a host of wildlife species.

**Creepy-crawlies: Friend or Foe**

Bat canines dripping with blood, snakes hiding in the shadows, maggots eating their way through decomposing flesh, ticks and mosquitoes mining our life fluids, worms and bacteria parasitizing our intestine—it all sounds like a bad dream. While many of these creepy-crawlies make our skin crawl, creatures in this category are in fact be the custodians of balance in ecosystems across the globe and they too have a story to tell.

Take the bats, for example. They are intelligent mammals that demonstrate some of the most evolved parenting skills and food-sharing observed in the animal kingdom. Many species of bats are so important to tree pollination that without their pollinating services, many fruits and nuts would simply disappear, affecting a whole host of wildlife including humans. Hoards of insects would descend upon us if it weren't for insectivorous bats. But bats are in danger of disappearing like so many other animals. In the eastern USA, more than a million bats have perished in the past several years from a skin fungus referred to as "White-nose Syndrome."

Known since biblical times as the spreader of sinful deeds, snakes, instead are bringers of balance. Without them, our world would be overrun with rodents (rats and mice) which would quickly consume our grain crops and spread disease that would put our lives at risk. Snakes are among the most highly engineered hunters on Earth! Because of their slender bodies, snakes are able to hunt in places like small burrows that would leave a cat panting for a rat. Research has clearly shown that where snakes are eliminated, human diseases like bubonic plague and hantavirus increase dramatically, in some cases, to epidemic proportions killing tens of thousands of people. Like bats, however, snakes are also in trouble. According to the Royal Society, snake populations worldwide are declining at unprecedented rates, largely due to gratuitous killing and habitat loss.

Macroparasites, which include helminths (worms) and some arthropods (ticks), can be much harder to appreciate. However, macroparasites are found in nearly every environment and healthy, thriving ecosystems have healthy macroparasite populations. Furthermore, macroparasites can provide information on population structure, evolutionary hypotheses, environmental stressors, trophic interactions, biodiversity, and climatic conditions which help us better understand our planet.

Some people are using a parasitic hookworm *Necator americanus,* to treat allergies like hay fever and asthma. When introduced in small numbers, symptoms of these human disorders are said to be reduced or eliminated. A new class of drugs that mimics the worm's effect could potentially treat celiac disease, autism, Crohn's disease, and arthritis.

Fly maggots can be used to eat (treat) human flesh that has become gangrenous and are the cleanup crews for rotting carcasses which keeps disease at bay. Medical leeches are employed to consume surplus blood that accumulates in amputated, reattached human fingers. Live garden slugs (*Arion hortensis*) are ingested to treat peptic ulcers and slime from the land snail (*Cornu aspersum*) is reported to cure warts and other skin disorders. The use of these creepy–crawlies is referred to as zootherapy.

Some research suggests that resistance or immunity to a substance requires early exposure. Studies in Europe report that children raised on farms were 30% to 50% less likely to have adult allergies than children raised in the city. According to scientists, farm-raised kids who were exposed to animals and pathogen-rich environments early in life developed stronger immune systems, while kids in the city, deficient in these early exposures, had higher levels of hypersensitivity.

It has been estimated that the average human is home to around 4 trillion viruses on or inside our bodies including as many as 500 species of bacteria in our intestines alone. Around 39 trillion microbes call our bodies home (around 4.5 pounds of our total body weight), compared to a mere 30 trillion human cells that make up our human forms. Our skin, bellybuttons, and hair are crawling with hundreds of species of microscopic organisms that would rival any metropolitan zoo! Without bacteria, our food absorption would cease to function properly, and without viruses consuming the surplus bacteria, we would be overrun with bacteria. Believe it or not, people who have lost these essential bacteria through the use of antibiotics are restoring the organisms into their bowels by ingesting the feces of healthy people—the results of which are reported to be astounding! If all viruses were to suddenly disappeared, life on earth would collapse—according to Tony Goldberg, an epidemiologist at the University of Wisconsin-Madison.

Rhinotillexomaina (the longest word in this book) is the compulsive act of picking one's nose. We all do it from time to time (although most of us use a handkerchief), cleaning out those unwanted accumulations. Mucophagy is the act of consuming nasal mucus accumulations (boogers, snot, nose crusts, bogey). Believe it or not, some individuals suggests that kids who eat their nose-candy may be healthier than kids that wipe their boogers under tabletops.

Boogers are full of bacteria that may protect teeth and be good for your gut, for the same reason probiotic foods like kimchi, sauerkraut and yogurt are touted as immune system boosters and even mood regulators. The Virus and bacteria we all breathe in during a walk in the woods, or in a crowd of people are trapped in the nasal mucus—forming antibodies, that might aid in fighting infections, acting like vaccines. But these are speculations at best and not back by scientific research. Socially unacceptable to most adults, gross to many but kids may just have it right! In any case, a book about biodiversity and discovery would seem incomplete without some discussion on what we consider the most obnoxious creatures on Earth.

*Creepy-crawlies, friend or foe? You decide!*

## Bioluminescence

Bioluminescence is natural light created by living organisms occurring in approximately 25 different phyla, many of which are entirely unrelated. These include arthropods, certain crustaceans, several species of mollusks including the land snail, *Quantula striata,* from Malaysia, deep -sea fishes, marine microorganisms known as dinoflagellates, fungi, and several varieties of slime molds. The most familiar animals in Belize possessing this unique power are the fireflies and headlight beetles. Light organs located on the abdomen and thorax regions of these insects are used to communicate and attract potential mates and in some cases, lure in prey to eat.

Several hypotheses of bioluminescence function in fungi include attracting hungry invertebrates which may aid in spore dispersal or attracting predators of mycetophagous invertebrates (organisms that eat mycelium). Some scientists speculate that glowing fungi are nothing more than by-products of some other metabolic process. In high elevation montane forests of the Maya Mountains, the jungle floor glows at night, a result of bioluminescent mycelium growing in decaying oak leaves and other decomposing vegetation. It is a sight to behold!

Biological lighting differs from other light sources such as light bulbs and candlelight in being cool, producing little or no heat during illumination. Bioluminescent light is the result of a biochemical reaction during which two types of chemicals luciferin and luciferase, are combined. The luciferase acts as an enzyme allowing the luciferin to release energy as it is oxidized, producing light. There is ongoing research involving bioluminescence in the areas of evolution, ecology, histology, physiology, biochemistry, and biomedical applications.

## Caves

Caves are unique habitats unlike anything found on the surface of the planet. They contain whole ecosystems in which species live, breed, and die without ever seeing the light of day. These creatures spend their entire lifetime in total darkness, so they are usually without pigment and functional eyes, and therefore referred to as troglobitic organisms. These are some of the most highly specialized animals found anywhere and include amblypygids, spiders, millipedes, gastropods, and crustaceans like the endemic Maya Mountain ghost crab. Most are extraordinarily rare, a result of isolation from surface environments. Troglobitic creatures are highly sensitive to changes in climate and human pollution. These organisms walk a thin line between life and death, entirely dependent on the waste material (guano) of crickets, bats, and other foods brought into the caves during flood events. It has been conservatively estimated that between twenty and thirty different taxa (many of the species new to science) occur in moderately-sized cave systems found in the Maya Mountains, most of which remain unstudied.

While little is known about troglobitic creatures and their diversity, caves are clearly key habitats for a number of well-known animals such as bats. Most larger systems will easily carry a dozen species of Chiroptera, sometimes numbering in the thousands. Not only are caves critical day-roosting sites, but caves also serve as night roosts for bats feeding on meals harvested from the jungle.

Caves are often referred to as living or dead systems, depending on their hydrology. Living caves are those caves that continue to grow in size (a result of flowing water) and ones that continue to cultivate stalactites, stalagmites, soda straws, columns, and flowstones.

When rainwater filters down through porous ground, a weak carbonic acid forms that eats away at the soft limestone. The solution, now enriched with dissolved limestone, finds an underground cavity and the mineral-rich waters leave behind minute deposits of calcium carbonate, creating and growing cave formations. In contrast, dead caves are systems without active hydrology but nearly all caves contain both active and inactive growth areas.

Deep within caves of the Maya Mountains live giants among their kind. They are some of the strangest animals on Earth, looking like something from a science fiction movie. These micro-beasts of the underworld are no less ferocious than *Tyrannosaurus rex* in the way they hunt, catch, and eat prey half alive. The ominous-looking Amblypygid or tailless whip-scorpion, (below image of this fantastic creature) uses modified spiny pedipalps for catching, holding, and devouring its prey of cave crickets and spiders.

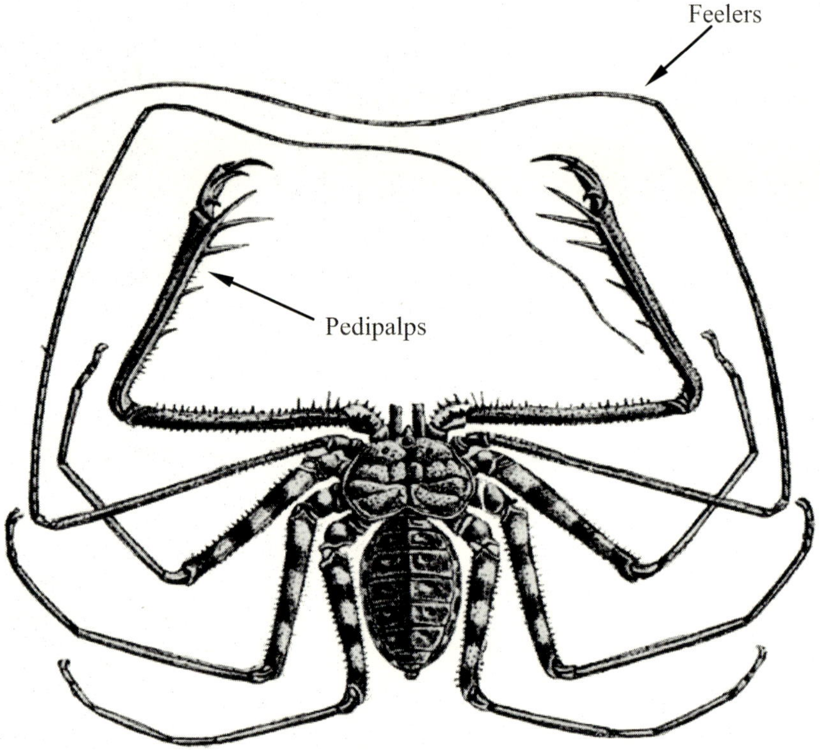


**Amblypygid (Actual Size)**

# New Land Snail Species From Belize Described by the Author, Caldwell & Thompson

**Ornate Crystal**
*Chanomphalus angelae*

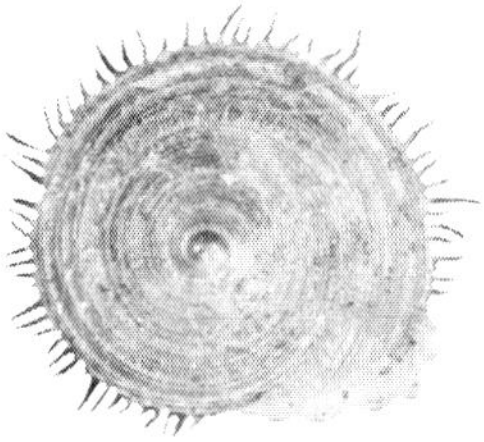

**Hairy Lucidella**
*Lucidella caldwelli*

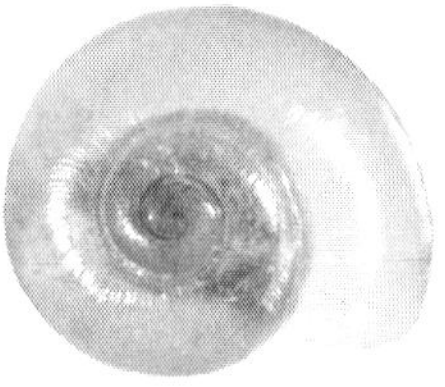

**Bladen Gloss**
*Miradiscops bladenensis*

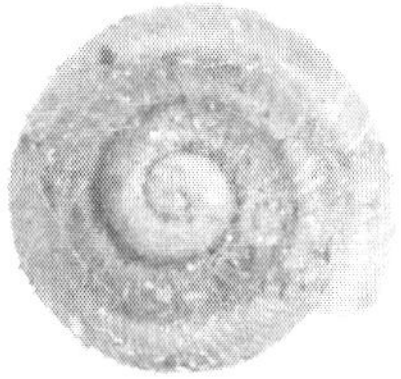

**Hairy Phora**
*Thysanophora meermani*

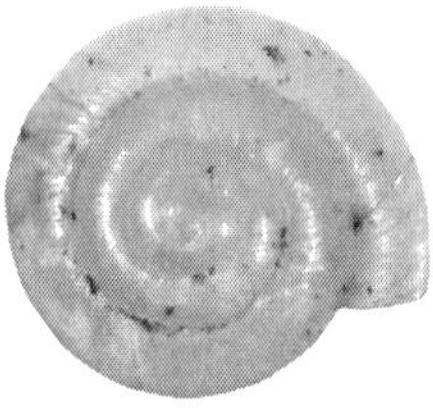

**Mountain Gloss**
*Miradiscops striatae*

**Hillside Gloss**
*Miradiscops youngii*

**Maya Mountain Rotadiscus**
*Rotadiscus saqui*

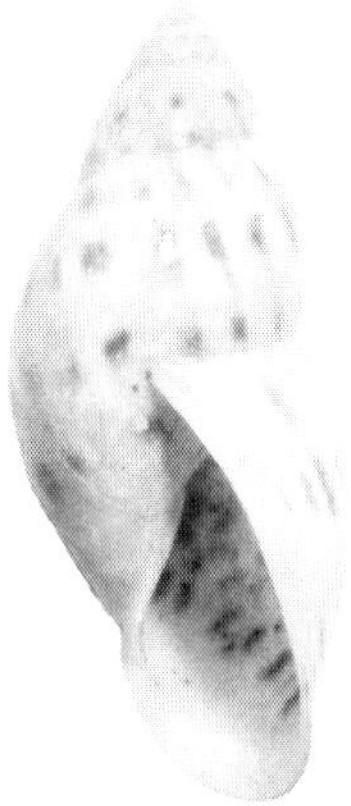

**Checkered Cone**
*Drymaeus tzubi*

**Oak Ridge Teardrop**
*Cecilioides dicaprio*

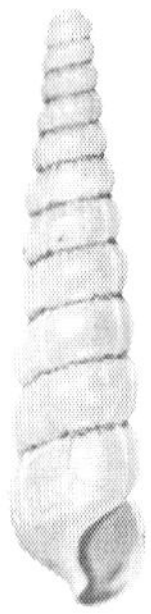

**Blue Creek Shaft**
*Pseudosubulina juancho*

**Smooth Quill**
*Brachypodella levisa*

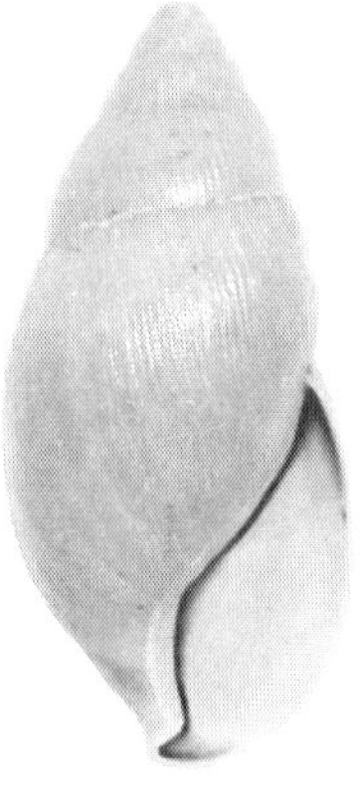

**Rosy Marauder**
*Euglandina fosteri*

**Montane Splinter**
*Rectaxis breweri*

**Mayan Cave Snail**
*Leptopeas corwinii*

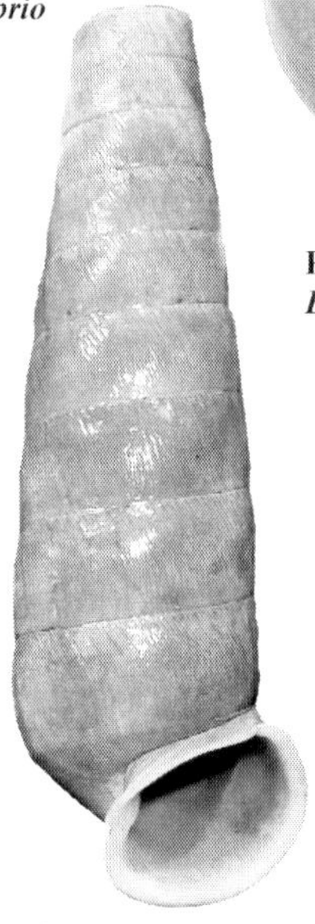

**Mayan Drum**
*Eucalodium belizensis*

**Blue Creek Cave Snail**
*Leptinaria doddi*

**Belize Cave Snail**
*Opeas marlini*

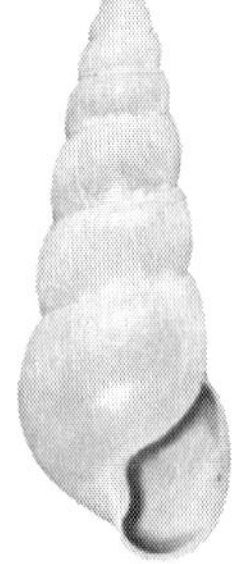

**Macal River Cave Snail**
*Lamellaxis matola*

# Beyond Belize

## Warning

## !Enter at Your Own Risk!

Past this point, the book digs into my thoughts about the state of the planet, where we are most likely headed, and the disastrous outcomes of our meager response to an ever-growing planetary crisis. It is not a pretty picture! It is sure to stir a rethink in some and no doubt hysteria in others. So be forewarned!

Writing this book also sent me careening down a difficult road of self-discovery, not only chronicling past fragments of my unconventional life—but providing understanding about why my life has been and continues to be—unconventional. I dare to share details about myself that until recently were even unknown to me. But there's still time to run and use the rest of the pages as TP and it would offend me not!

# My Past Environmental Views
## *Sidestepped Truths of Yesterday*

For a moment, imagine the ride of a lifetime, full of mystery and adventure around every turn. Throughout the ride, you discover treasures beyond your wildest dreams. Bursting with excitement, your senses are on overload, and best of all, it's free.

The ride? Spaceship Earth, a wonderfully time-crafted vessel of unimaginable beauty hanging in the balance. The treasures discovered; life itself amid sapphire-blue seas and distant mountains shrouded in mist, teeming with a rich tapestry of plants and animals.

This is Belize today. Now imagine your children taking this same ride as adults, but seeing ever-increasing polluted waterways or palm-oil plantations where immense herds of warrie were once hunted by magnificent spotted cats. This precarious ride has quickly become the fate of much of Central and South America, notwithstanding the rest of the world. We are now living in the 7th mass extinction of our planet and, for the first time, it is not driven by the geophysical forces of the planet or from beyond, but coming from its own creation—*Homo sapiens*.

Ernst Mayr considered the father of modern biology, proposed the contentious idea that human intelligence was a "lethal mutation". He may be correct if, in our quest for a so-called better life, we leave a wake of destruction where elephants no longer roam the savannas of Africa, monarch butterflies are but distant memories, and for our children and grandkids, a tiger is nothing more than a bedtime story. As we continue to grind the planet to dust, these events are no longer predictions, but realities in many parts of the world, and it's only getting worse.

Given the state of the environment, I often ponder the future of the planet and the quandary of our species. What were once pristine rivers of vitality are now gutters of spoil, oceans of prosperity turned to waters of poverty, and forests untainted by human activities are becoming but distant memories. Our world is not infinite but a place of boundaries. Clearly, our species has pushed many of these boundaries and beyond. Our ever-increasing demand for Earth's limited resources has led us down a dangerous pathway of no return, tearing away the delicate strata of life on which we all depend. In an age of enlightenment, why are we still losing ground on which we stand? Many fault corporate and government entities, but blame does not fall on the heads of a few. Nevertheless if you feel an obligatory need to point a finger, make sure it's got a U-turn in it!!!

In the last century, we left our agrarian lifestyle behind, trading our kinship to the land for a lifeless relationship with technology. It is a perplexing time for us all, living in a world of machines, but still entirely dependent on the environment. Spending much of our time trapped behind walls of concrete and steel, we have become desensitized to a planet in peril. While we may be better aware of the plight, our species turns a blind eye and continues on a reckless course of consumerism, exhausting the planet's finite resources. The transfer of this uncertainty to our children has far worse implications. According to Richard Louv, author of "Last Child in the Woods," a fine book about our human disconnect to the natural world, the lack of connectivity to planet Earth has created a generation that may well be suffering a non-medical condition aptly labeled "nature-deficit disorder".

We rape and take until little remains for the future and in our last dying breath and golden opportunity to do the right thing, we fail miserably. After decades of consuming life-giving resources our body dies, is drained of its fluids, replaced by impious chemicals, buried in isolation in ridiculously ornamented boxes, neither body or casket is ever seen again. What a bizarre and wasteful use of precious and finite resources! This last great insult is to truly rot in hell, denying our corpses obligation to return its essence to whence it came. No other creature on Earth behaves in this irresponsible manner. Besides the Bible commands it of us in Genesis 3:19, "Ashes to ashes and dust to dust" we shall return. The Bible got this passage right! A better way of dealing with our dead might look something like this with our bodies buried in biodegradable linens under a planted tree, beginning to heal the only world we have.

Environmental ethics is not a way to imagine. It is a way of life; a life that honors Earth and its inhabitants; a life that considers no less than the whole. The decisions we make every day, singly or collectively, have far-reaching and sometimes irreversible consequences to the land; land that our progeny and fellow creatures will inherit. If we are to protect the things around us, we must all think in terms of how our every action, from flushing our toilets to the foods we eat, affects our planet and the life force within it. **Becoming better stewards of the land, air, and water makes sense, regardless of climate change**. Caring for the environment is everybody's responsibility. If we fail as competent caretakers of our native soil, all that follows, including our children, will surely suffer the consequences.

***"We live among relatives, not resources." ~Daniel Wildcat***

# My Current Environmental Views
## *Inconvenient Truths of Tomorrow*

***Earth creates Nature***
***Nature gives rise to Man***
***Man destroys Nature***

*The Earth has been diagnosed with stage four cancer. The cause is Homo siphons and the prognosis is terminal.*

By now you are probably thinking the environmental conflict within me is about as clear as the muddy waters of the Amazon. The conundrum is simple; my past desires to make a genuine difference with regards to the environment and the current realization that the world is just basically fucked. As a biologist working at ground zero, I have witnessed profound changes in mere decades. One of the saddest and yet environmentally sobering moments in my life was to watch bats I had spent my career protecting, fall like weeds in the wind. Seeing these remarkable animals disappear before my very eyes was heart-wrenching. I know now that my work in conservation had little to do with saving wildlife and more to do with documenting their demise. No longer an eternal optimist, my tears have run dry for an ever-disappearing wild.

*Remove a few cogs of a wheel it begins to wobble,*
*remove many cogs, it falls apart.*

Nowadays I consider myself nothing more than a **Biota-historian**, writing books about an unwelcomed future—the books serving as an archive of pictures of what once was and a painful reminder of our indifference to a vanishing wilderness and its wildlife; all lost for what we were told and believed would be a better life. But for who? Certainly not for the nearly 700 vertebrate species we have already driven to extinction since the 16th century and the millions predicted to be gone by the end of 2100. Consequently, my "saving the planet" illusion bubble burst many years ago.

I was not always this way. There was a time I actually believed that humanity would wake up and see the dying roses. But it never did, not really, and so time marched/marches onward, ever-consuming what little of Earth resources remain. One of the downsides of being so damn open-minded is that you rarely close the gates to new information, which at times, strands the mind on an ever-sinking ship of despair. For me, it was the overwhelming and depressing details of increasing environmental problems across the globe with no determined effort to fix it.

I used to think that when the grass turned brown and we choked on the air we breathe, humanity would finally rouse from its slumber. Although both are now realities, we continue as if nothing is wrong! Ever-expanding globalization, technology, and uncontrolled human population all striving for a better life, doesn't come without a profound toll on a planet that barely hovers in the Goldilocks Zone of the universe.

Research by Johan Rockström (founding director of Sweden's Stockholm Resilience Centre) and 28 other internationally renowned scientists concluded that there are "9 planetary boundaries" (seen below and page 176) humanity must not cross to keep the planet habitable for *Homo sapiens,* not to mention the rest of life on earth!. Six have already been crossed, one close to being crossed, one not yet quantified, and only one that was actually reversed through the ban of chlorofluorocarbons (CFCs) and halons—gases formerly found in aerosol spray cans and refrigerants that degraded the ozone layer of the atmosphere. But CFCs are still in use in some countries and in 2020 and 2021 the ozone hole once again has experienced an increase in size.

- **Climate change** (already crossed)
- **Biosphere Integrity-Biodiversity** (already crossed)
- **Ocean acidification** (close to being crossed)
- **Freshwater use** (crossed in 2022)
- **Ozone depletion** (crossed but reversed in the 1990s)
- **Atmospheric aerosol pollution** (not yet quantified)
- **Biogeochemical flows of N and P** (already crossed)
- **Land-system change** (already crossed)
- **Release of novel chemicals** (crossed in 2022)

The core of Rockström research is well summarized by Claire Asher in an article written for Mongabay, March 30, 2021. She writes,

> "All life on Earth, and human civilization, are sustained by vital biogeochemical systems, which are in delicate balance. However, our species—due largely to rapid population growth and explosive **consumption**—is destabilizing these Earth processes, endangering the stability of the "safe operating space for humanity".

Since six of the nine Earth Systems are already beyond a safe position, we are careening toward 'Hothouse Earth' (illustrated below), from which it would be difficult, if not impossible to recover to a pre-industrial climate.

## The Holocene Epoch

Approximately 12,000 years ago marked the end of an unstable and fluctuating glacial and interglacial era where Earth temperatures varied by as much as 6°C (10.8°F). The Holocene Epoch followed (see graph top of next page), with a relativity stable climate where *Homo sapiens* have been evolving civilization ever since. This included the domestication of animals and plants as well as the establishment of agriculture, allowing our species to flourish. We, however, are now pushing the Earth to a new era, the "Anthropogenic" (illustrated below).

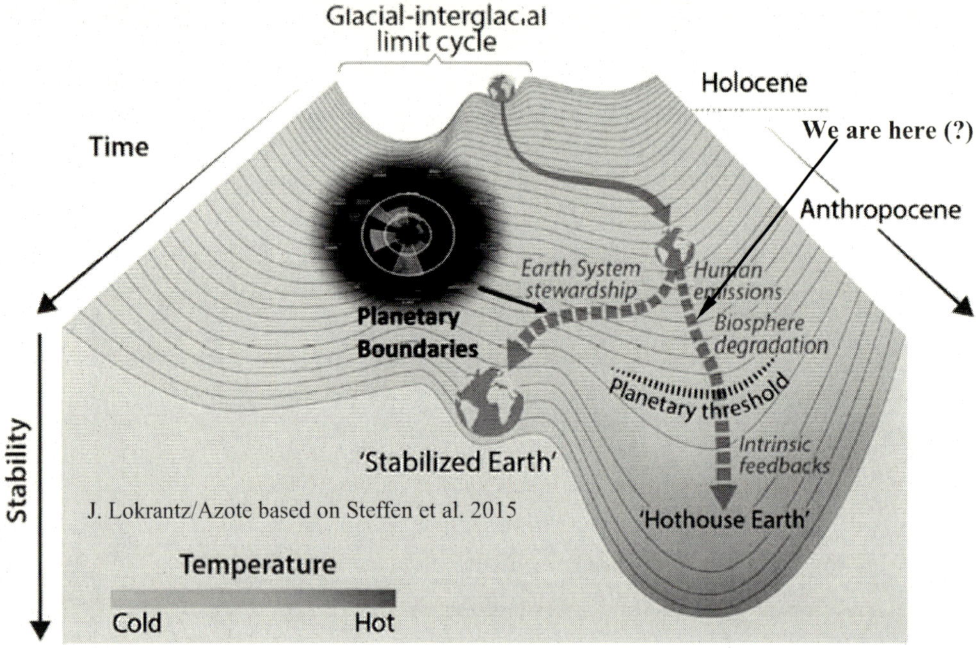


Steffen et al. 2018

## The Anthropogenic Epoch

Sometime along the Holocene epoch timeline, humans began to change the Earth in meaningful ways by human-triggered increases in greenhouse gases, large-scale land-use, excessive biodiversity loss, and massive global consumption, and pollution brought on by ever advancing technology and fast-growing population. This newest time period has been dubbed the "Anthropogenic Epoch" (see graph above). Rockström suggested this new era is where humanity begins to hit the ceiling of the biophysical coping capacity of the whole Earth System, further adding

## 100,000 Years of Climate Change

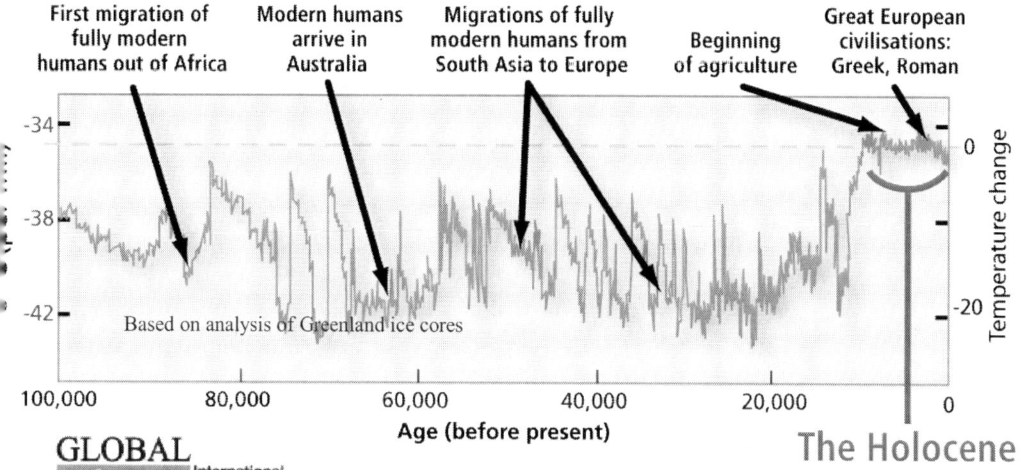


GLOBAL IGBP CHANGE International Geosphere-Biosphere Programme

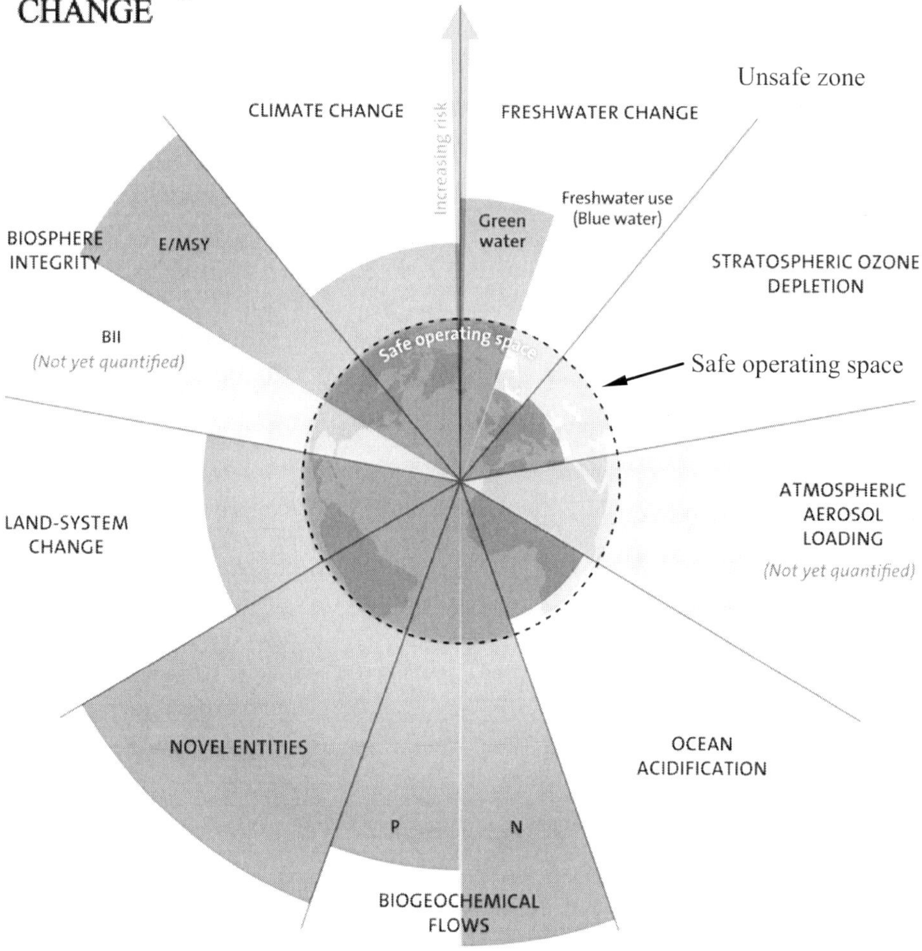


Source: Wang-Erlandsson et al (2022)

that there is little evidence we've reversed course to avoid ominous tipping points. If anything, we are even deeper into the transgression of climate, biodiversity, land-use with nitrogen and phosphorus imbalances.

Of the six boundaries that we have already exceeded, Rockström considers **climate change** and **biosphere integrity** to be "core" planetary boundaries because either one, on its own, could change the course of Earth's trajectory and endanger not only humanity but all other complex life forms. There's mountains of evidence for human-induced climate change and the 7th mass extinction event is well on its way to pushing the planet ever-closer to "GAME OVER".

Curiously, research conduced at the Stockholm Resilience Centre suggests that Earth could feed 10 billion people if we stay within the 4 interlinked planetary boundaries; biosphere integrity (already crossed), nitrogen & phosphorus flows (already crossed), land-system change (already crossed), and freshwater use (crossed in 2022), but as such, would require radically different ways of farming, reduction of food waste, and dietary changes. Reaching this level of human density would require a major shift away from red meat and a move toward a vegetarian/vegan diet, as just one of many examples stated by the research.

Even "if" we could stay within the four planetary boundaries which have already been crossed with regards to our food production, it ignores one important but often overlooked human disparity—an equitable and productive life for all human beings. After all, just feeding people to keep them alive differs little from a feed lot. To put things in perspective—consider that each American uses around 9.7 hectares of land to support their consumptive lifestyles. This means that Earth can support 1.5 billion people, living an average American lifestyle. But to bring the current population (7.5 billion people) up to the standard of living that most Americans enjoy today, an additional five and a half Earths would be needed according to the Global Footprint Network!

Ozzie Zehner who wrote *Green Illusions, The Dirty Secrets of Clean Energy and the Future of Environmentalism*, asserts "We don't have an energy crisis, we have a consumption crisis"; although I would argue that they are one in the same.

The UN Development Program (UNDP) reported the richest 20 percent of the world's population consumes 86% of the planet's resources while the poorest 80% uses just 14 percent. The US makes up only 5% of the world population yet uses 24% of the world's energy, produces more waste than any other country, and every year, throws away 108 billion pounds (40 percent) of our food into landfills, worth an estimated $161 billion, and yet there remains a hunger crisis in the US and across the globe.

---

Putting the consumption issues aside for a moment, and asking the million dollar question "***Is green renewable energy a bona fide fix to our energy crisis?***"

---

While the idea of clean renewable technologies such as the sun, wind, and hydro is alluring, are they just feel good solutions that aren't solutions at all, being promoted to soothe the guilt we feel for trashing the planet? Whereas, the downsides of coal, oil, and gas are well documented, the dark side of renewables is ambiguous at best, especially biofuels.

Taylor C. Dotson writes, "The framing of PVs and wind turbines as "green," "renewable," and "zero-carbon" distracts us from how *all* energy technologies lead to deaths (animal and human), ecosystem destruction, massive levels of extraction and processing of raw materials, pollution, and even the disruption of our experience of non-human spaces".

As just a few examples of the dirty side of green energy, consider that windmills use vast quantities of neodymium in their magnets, which is mined like coal, using fossil fuels, not to mention animal mortality.

According to the US Fish and Wildlife Service, around 888,000 bats and between 140,000 and 500,000 birds are killed every year by wind farms. Although bats can detect and avoid the rotating blades, they often suffer from barotrauma—an internal ear injury caused by exposure to rapid pressure changes near the trailing edges of the blades, usually lethal to bats. White-nose Syndrome has already killed millions of bats and wind farms may be the final death sentence for several species!

Solar panels require rare earth metals such as yttrium or europium that are also mined from the Earth with heavy machinery that runs on diesel. Hydroelectric damns turn richly diverse river ecosystems into lakes that support far less biodiversity. There really is no such thing as a free lunch, even with renewables.

Other renewable sources include biomass (trees, switch grass, corn, sugarcane, and algae) used to produce power and biofuels (ethane).

Similar to solar, wind, and hydro, biomass also requires fossil fuels in both transportation and production, with their carbon footprint and cost-benefits remaining vague at best.

As one example of biofuels, forests are cut down and turned into wood pellets to produce power. One study estimated that wood pellets made in the U.S. and burned in the U.K. led to 13-16 million tons of $CO_2$ emissions in 2019 alone, equivalent to the emissions of up to 7 million cars. Wood pellets are, in fact, less energy-dense than coal and produce more carbon pollution per unit burned than coal does to generate the same amount of electricity. Furthermore, there can be far-reaching impacts on biodiversity, depending on the type of tree-harvest methods employed (i.e. clear cutting versus select cutting).

Consequently, the use of biomass to produce energy remains highly controversial. Clearly, all sources of green energy are still *currently and undeniably* tied and dependent on the use of oil, gas, or coal in their production.

---

Putting the energy production issues aside and asking the other million dollar question, "***Is consumption the bigger issue we need to be addressing***"?

---

## *Unequivocally YES!*

To put it bluntly, infinite consumption of finite resources is not sustainable and will only lead us all to premature planetary hospice. Furthermore, "*Sustainable Development*", an idea put forth by leaders and politicians, is a cowardly and self-serving platform—that drives the ever-increasing consequence—of ever-decreasing resources.

Rethinking the phrase might look something like
"Capacity of Sustainability" instead of "Sustainable Development"

In other words, what is the carrying capacity or limits of a particular area, taking into consideration water, land, and the needs of wildlife? After all, we are obliterating their homes when we build a parking lot or began a new subdivision. How far should a town grow and expand its boundaries before someone says "no more", words a politician would never think, let alone dare speak? "Sustainable Development" could be characterized as "hypocrisy without end".

The question we should all be asking ourselves is,
"*When is a enough—enough*???"

Richard Heinberg, author of "*Power, Limits of Prospects for Human Survival*" had this to say, "We now move more of the planet's crust, just in the process of mining for minerals, farming, and building than is moved by all-natural processes combined (i.e. wind, rain, hurricanes, earthquakes)". Over-consumptive factors such as energy, housing, and the endless manufacturing demands for technological devices are the drivers of our relentless disembowelment of the Earth.

To put things in perspective, consider cement as just one of many examples of over-consumption. Seems innocent enough, right? But for every ton of cement made, a ton of $CO_2$ is released into the atmosphere and the industry generates around 2.8 billion metric tons of it every year. That's 8% of the total global $CO_2$ emissions to erect our never-ending buildings, roads, bridges, and other infrastructure.

*The ones that produce the power we love, are the same ones that wield the power we hate.*

**The Conundrum We All Now Face**

Unfortunately, humanity is now one big tangled mess, and tugging at any one strand disrupts its imaginary continuity, causing even more disarray somewhere, or to someone. This is why governments are slow to change anything that is to the contrary, and why the Kyoto Protocol (2005) and Paris Agreement (2015) have failed to gain any notable traction of consequence. Since their conception and signing thereof, carbon emissions have only continued to rise—from 382 parts per million (ppm) in 2005 to their current 415 (ppm). In 1950, it was around 300 (ppm), and around 283 (ppm) in the 1800s. At the COP26 Climate Talks in 2021, the theme was more Blah-Blah-Blah and business as usual, failing once again to address the looming catastrophe.

*If you think that the environment is less important than the economy, try holding your breath while you count your money.* ~Guy McPherson

Further damning information with regards to our current crisis includes the most recent IPCC (Intergovernmental Panel on Climate Change) report (one of six) stating a "**Code Red for Humanity**". The alarm bells are deafening, and the evidence of a warming planet is irrefutable". The IPCC warns that sea levels are and will continue to rise (as much as 6 feet by 2100), storms, and droughts will only intensify, and forests will burn to the ground. We are currently on track for a rise of 5° C or greater by the end of the century, with consequences that will look worse than a post-apocalyptic Mad-Max movie. Wet-bulb temperatures are on the rise and at 32 °C (90 °F), equivalent to a heat index of 55 °C

(130 °F), humans and other complex organisms loose the ability to thermal regulate and die. Even if we could solve the energy problem with renewables, the ever-increasing population of *Homo siphons* would continue siphoning and tearing open the planet for its finite resources and its precious wildlife. Fantastic creatures that no longer roam the Earth with **us**, as a result of being eaten to extinction by **us**, include the Dodo Bird, Steller's Sea Cow, Passenger Pigeon, Caribbean Monk Seal, Wooly Mammoth, and the Great Auk, as just a few examples.

Viewed by many as just another resource to be exploited, or a source of needed protein, animals continue to be sold in the pet trade or eaten at unprecedented rates. And no species is safe from human consumption, including chimpanzees and gorillas, which are currently being murdered in Africa for bushmeat. This is a problem on several levels including evidence that strongly suggests that wild meat consumption is linked to "Zoonotic Diseases" such as monkeypox, SARS, Sudan, and Zaire Ebola, and perhaps COVID-19. Furthermore, human activities have caused the world's wildlife populations (both rare and common species) to plummet by more than two-thirds (68%) in the last 50 years, according to a recent 2020 report from the World Wildlife Fund. It's simple mathematics. *The more of us there are, the less of them remain!*

According to the climate experts, if we are to save a livable planet for our children, we need to stop all carbon emission today (not tomorrow), to prevent a runaway climate catastrophe. *How do we quite emitting* $CO_2$ *gases today*?

While the answer is quite simple, the action needed is virtually unattainable. WHY? Because it would require, but not be limited to, turning everything off, stop driving cars, flying, stop cutting trees, eating differently, living in smaller houses, and raising a garden—in other words—stop living like the 20% and start living like the 80%. Those of us with more need to learn to live with less, so those with less can live with more. But we all know that ain't going to happen. What's more, turning everything off will actually make things worse, heating the planet even faster, a result of the **Aerosol Masking Effect**. The burning of fossil fuels has put particulates in the atmosphere that has a dimming effect, keeping the planet cooler. Caught between a rock and a hard place!!!

*We "**speak**" as if we are worried about our kids' future,*
*we "**act**" as if we don't care.*

Let's face it, the animal, *Homo sapiens* is caught between two worlds. One where we put our survival over all other creatures and the other where we feel empathy for those same creatures we crush in our path.

It's a tug-of-war we are losing—where one man celebrates the death of the last great blue whale, even as another is giving his life to save it. No other animal on Earth is trapped by these internal conflicts. And while humans can demonstrate incredible feats of kindness, it's only to the point where it doesn't compromise our consumptive lifestyles. Without question, Americans are among the most privileged species to ever walk the Earth! So be grateful for what you have—not what you don't!

Our survival, as with all species, will always be first and foremost to benefit *Homo siphons*! Don't believe me? In July of 2021, the United Nations released a draft of the Post-2020 Global Biodiversity Framework, which called for 30% of Earth's land and sea areas to be conserved—a planet that sustains 8.7 million plant and animal species (as much as 80% still waiting discovery). The idea of keeping 70 percent of Earth for ourselves (one species, along with a few pets, livestock, and food crops), leaving only 30 percent for the rest of the planet (around 9 million species excluding the yet-to-be found ones), is troubling to say the least. Perhaps worrying over percentages is a moot point. According to Guy McPherson, one of the first persons to openly connect the dots of numerous feedback loops had this to say, "*We are on the verge of losing habitat for Homo sapiens.*" (Nature Bats Last; April 30, 2021). It begs the question, can we ever be more than a collection of impetuous animals—unable to break free of instinctive, self-serving aspirations?

I often hear "if we just act now, better late than never, right?" Unfortunately, that better late than never was 30 years ago when in 1988, James Hansen, a former NASA scientist uttered these words as he sat before Congress. "*The greenhouse effect has been detected, and it is changing our climate now.*" Even Exxon scientists knew about climate change 40 years ago, but buried their research and began a propaganda campaign to convince us all that climate change was a hoax. It worked like a charm. Sadly, our government and we, the people did nothing then, and are doing little more than talking about it now! No doubt governments have already considered the information presented here, and as such, adopted the policy that "*business as usual*" is really all we can do. As hard as this is for me to say, I don't entirely disagree.

**Ignorance** is bliss, a blanket most will hide under until the bitter end. **Denial** of science doesn't alter the facts. Sticking our heads in the sand won't change environmental degradation any more than ignoring cancer. Both slowly eat away the subtle fibers of life we all depend on. **Hope** and **Fear** are exploited by those who look to control and keep the masses in check and from taking right action. **Hopeium** is held by those thinking that something (God?) or someone (not me?) will fix the troubled world. Hope without action clearly gets us nowhere.

And then there's just good old-fashioned **lying** that seems intrinsic to our species. The reason we find solace in lies is that dishonesty is often less painful than honesty.

Under the auspices of not wanting to hurt other's feelings, humans would rather be told what they WANT to hear, not what you ACTUALLY THINK, which is essentially saying humanity can't deal with the truth. But, in my opinion, telling truth is almost always better than hiding truth. Or to quote James Walton, "*Hypocrisy in search of social acceptance erodes your self-respect*".

We have now become a society comfortable with lying to ourselves, to our friends, our family, even our kids. Moreover, falsehoods allow us to remain unchanging and guilt-free from the uncomfortable truths of the past and the looming truths of tomorrow. Telling half-truths is even worse, keeping people in the dark and from making informed decisions for lack of all the information.

Nevertheless, at some point, I decided to let the **inconvenient truths of tomorrow** go, and would no longer be "dragged" or "attached" to any outcomes. It's only when you send "Denial and Hopeium" adrift that one can reach unfiltered conclusions, however, painful or troubling they might be. How I got here is anyone's guess, but a few realizations came to my Dyslexic and Asperger's brain and are as follows.

Great defenders of the Earth, such as Aldo Leopold, John Muir, and Rachel Carson have come and gone. Hundreds of "how to fix the problem" books have been written, read, and shelved. Dozens of celebrated environmental documentaries have been made, watched, and forgotten, and yet, we continue to edge ever-closer to near-term human extinction and a planet on fire. Our collective short attention spans on important issues such as **Anthropocene Extinction**, and our long-attention spans on trivial issues like the latest Kardashian Drama will ultimately bring us to that inevitable question, "Why didn't we listen to the experts and act sooner"? Sadly that sooner is now 30 years past!

It's bizarre to me that we spend ridiculous amounts of time and money training great minds in physics, science, and mathematics to help solve the looming problems of the world, yet turn around and poke fun at these same experts and their science, while bolstering the stupid and the ignorant.

*Human intelligence is our nemesis, where the realm of careless thought has no boundaries.*

Although we rarely question the science when it comes to our health (even this is changing), when it comes to the science of climate change, it's denial we take to our grave. Why? Because it is much easier to pop a pill than to change a way of life. Realizing that there's a problem and how to fix it isn't humanity's biggest obstacle anymore, it is the collective will to act on it.

And then religions would have us all believe that some magic person in the sky is going to fix everything, no matter how bad we fuck it up. So what's the incentive to put right our mistakes with that kind of childish thinking? Religions will always be a barrier to advancing scientifically-based agendas.

You could assemble the best scientific minds, task them with the most expeditious way to tear down the Earth's last fragments of life-giving resources and they would fall shy of the planet's current and future catastrophic trajectory. No matter how we power and feed human civilization, it's an ever-growing, exponential heat engine—a conundrum I suspect "science" will be unable to fix in time. Even as our species takes its last breaths of stagnant air, the Earth will morn not our passing any more than the death of a virus or the billion other creatures that have come and gone. And the universe will continue on its impartial journey until its failure and inescapable heat death.

If Ernst Mayr was right about human intelligence being a lethal mutation, then the species, *Homo siphon,* was not the pinnacle of human evolution, but the pinnacle of its demise.

**So is there a rational solution to the conundrum we now face?**

**Yodel-for Joy**

**Embrace-Life**

**Smile-you are Loved**

In our fleeting moments in time, seek out your passions while abandoning your indifferences. Be grateful for that astonishing endowment of life. Find solace and humble kinship with Earth's remaining wonders. Embrace sincere friendships and family with passion, for all we ever have with certainty are the moments!

# Perspectives on Human Civilization

## Ecological Overshoot

"Ecological Overshoot occurs when human demand exceeds the regenerative capacity of a natural ecosystem. Global Overshoot occurs when humanity demands more than what the biosphere can renew. In other words, humanity's Ecological Footprint exceeds what the planet can regenerate. The biosphere's renewable capacity includes the replenishment of resources and the absorption of waste, such as carbon dioxide from fossil fuel." ~ earthovershootday.org

*"Our core ecological problem is not climate change. It is overshoot, of which global warming is a symptom."* ~ Richard Heinberg

## Carrying Capacity

"*Carrying capacity can be defined as maximum population size that an ecosystem can support without being degraded in some fashion.*"~ Tom Wessels, The Myth of Progress.

Based on the two concepts stated above, we have already exceeded both Ecological Overshoot and Carrying Capacity of *Homo sapiens* on this planet!! While most people will deny the evidence, especially those who benefit most from the existing set of living arrangements, it doesn't change the inevitable and undeniable correlation between increasing civilization and decreasing biosphere integrity. *"Civilization is, by its very nature, a long-running* ***Ponzi Scheme****. It lives by robbing nature and borrowing from the future, exploiting its hinterland until there is nothing left to exploit, after which it implodes. While it still lives, it generates a temporary and fictitious surplus that it uses to enrich and empower the few and to dispossess and dominate the many."* ~ William Ophuls.

Historians, anthropologists, and others have proposed various explanations for why civilizations fail including: local weather events, environmental degradation, inequality and oligarchy, growing complexity, and external shocks (war, natural disasters, famine and plagues). Unsurprisingly, as societies grow in size and prosper, overshoot of resources and collapse of ecosystems are not far behind. In the past 10,000 years, HUMAN WELL-BEING for the privileged few has never been better and for the BIOSPHERE, it's never been worse. But the most consequential difference between failed civilizations in the last 10,000 years and our current one is globalization—a worldwide conduit that has spread diseases, invasive biota, and anthropogenic destruction to every corner of the planet. In the past, when a civilization collapsed, the land, given enough time, could heal itself thanks to nature's built-in resilience. But nature back then wasn't imbedded simultaneously with overly amplified climate chaos, melting glaciers, ocean acidification and heating, ozone depletion, atmospheric aerosol pollution, nitrogen and phosphorus over-loading, the insect apocalypse, and global biodiversity declines like it is today. And when our civilization finally implodes, there's one final nail we will place in Earth's coffin—441 nuclear power plants and their in-evitable meltdowns—smothering the planet in ionizing radiation—while stripping away the protective ozone layer for a millennium!

*"Posterity doesn't vote, and doesn't exert much influence in the market-place. So the living go on stealing from their descendants."* ~ William R. Catton excerpted from Overshoot, The Ecological Basis of Revolutionary Change

While most are oblivious or dismissive to the looming problems we now face, nearly all lack understanding of ecological overshoot and limits of carrying capacity. Those that do—cling to hopieum—believing that techno -fixes will somehow prevent collapse. Nevertheless, it remains doubtful that our problems can be solved with more of the same economic growth, technology, or questionable green energy. Our predicament weighs heavily on civilization; a state we were born into whether we like it or not! While civilization is largely what we want to save—it seems predestined to bring us all down! The paradox is that our solutions, thus far, have been solutions to preserve civilization. "*But insanity is doing the same thing over and over and expecting different results."* ~ Albert Einstein. At the cross-roads of planetary catastrophe, we need courage, compassion, and love not denial, hopeium, or more lies to deal with our impending calamity. Why? Because our next and final destination on spaceship Earth is ap-proaching widespread human/biota annihilation.

*"The end of the human race will be that it will eventually die of civilization."*~ Ralph Waldo Emerson

# My Life's Convoluted Journey

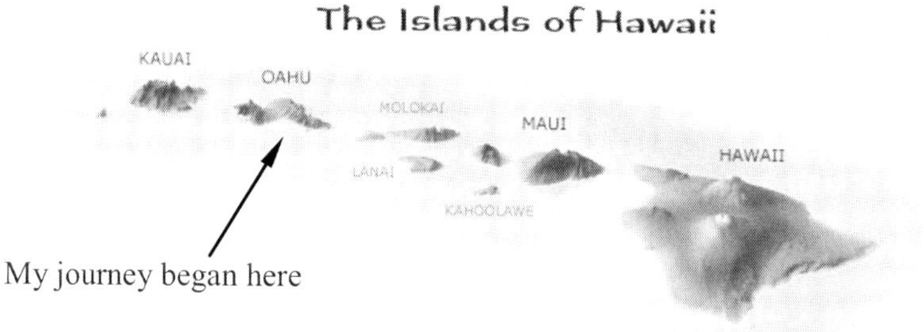


My first breaths of life were of warm tropical air on the island of Oahu, in the small ocean town of Aiea, Hawaii. I crawled out of my mother's womb in an Army Hospital at Pearl Harbor where my Navy dad was stationed at the time. This colorblind or "non-white vision impaired" (to be politically correct) island boy from Oahu would forever long to return to a sun-drenched and sultry environment. Little did I know it would take me 52 years!

In the interim, my life, indeed my passion, would center around the natural world. In early childhood, chasing down and catching bugs was my thing! In fact, according to my mother, my first spoken words, at age five, were "bad bug", vocabulary she often uttered in response to my bug-filled jars. At the time, I likely took that to mean "good bug"—little did I know. Nevertheless, I was allowed to keep all kinds of wild animals, from a resident raccoon named "Sparky" that terrorized my mother's friends during tea parties to an escaped black rat snake that laid 10 eggs in my chest-of-drawers. I knew the missing serpent had been discovered when my panic-stricken mom shrieked my name in concert with words I had never heard from my mother's mouth.

The neighborhood insurgent was Sparky the raccoon, which would routinely browbeat dogs into submission and chase pet cats up trees. Sparky also got into my neighbor's house one time, but not before walking across their freshly painted deck. You can only imagine how that turned out. As my animal repertoire increased, so did my mom's growing frustration with my pets having free reign of the house. But I hated the idea of caging my wild friends. Nevertheless, these blissful days were coming to an end as my mother's patience wore thin with each new animal addition. She would soon reach a tipping point.

It all happened one spring day when my loveable, but somewhat unruly pet pigeon, aptly named Pige, decided to crash another one of my mom's outdoor parties. For whatever reason, Pige set her sights on one of my mom's more annoying friends, Joan. Swooping down from a tree branch, Pige landed directly on Joan's head which sent her running and screaming all kinds of profanity, never to return. That day was especially memorable for people and animals alike. My mother had finally snapped, unleashing her full fury not upon me, but on my father, who otherwise supported my collection of disorderly critters. Unfortunately, that would be the day my pets were to be imprisoned or released back into the wild. It was indeed a sad day for all except perhaps my mother. For me, these precious wild animal experiences cultivated close bonds with something other than *Homo sapiens,* but for my mom, no doubt a belated heart condition. Sorry, Mom!

My first and most memorable employment was picking peppers, along with my brothers Mike and David, on a pepper farm in rural Cincinnati for fifty cents an hour, plus all the peppers you could eat. But let's face it, how many peppers would kids want to eat after picking peppers all day long? The answer of course is none!! But how many peppers would pepper-pickers throw at passing cars while in the back of the pepper truck is nearly as many peppers as pepper-pickers picked that day? As you might imagine that did not end well for my brothers and me, the "peppy-pepper-pickers." Try saying that three times fast!

Other early years of employment would include becoming a soda-jerk, scooping ice-cream at my mom and dad's ice-cream parlor in Cincinnati, followed by working as an EMT/paramedic on a life squad and Emergency Room Technician in a hospital, a machinist, a mechanical draftsman, and a timber crew member planting trees and fighting forest fires both locally and in western states. For a short time, I was a law enforcement officer with the US Forest Service which included training at the Federal Law Enforcement Training Center (FLETC), Glynco, Georgia with border patrol, FBI, and Secret Service agents. It was a wild ride! Cutting marijuana and busting poor Appalachian folks, though—would be my shamefaced days with the US Forest Service! But my most important achievement in terms of a turning point in my life would be a lengthy walkabout in California. As time marched on, I found myself increasingly yearning for a career direction. It wasn't until I walked the Pacific Crest Trail that I would take time to ponder a livelihood. Nearly six months into the trail and seventeen hundred miles later, I had come to a decision—a profession in the great outdoors.

## The Ecstasy of Dyslexia

The idea of college at age 28 and having Dyslexia with a side order of Auditory Processing Disorder (APD), was daunting, to say the least—anxiety born from my absolutely miserable days spent in grades K-12, or as I like to say "the incarcerated years." Having dyslexia and APD brought me much anguish as an adolescent that continued into adulthood. During my school years, my grades were never above Cs, mostly Ds, and often Fs, and the ridicule I received from peers and teachers was, at times downright brutal. In fact, I had to repeat second grade because my grades were so bad. One of my delightful "Catholic nun-educators", Sister Rose would tell me, "Dan, you're either not studying or you're just too stupid to understand the subject matter." The fact that I was indeed studying left me with only one tormenting conclusion. I was dim-witted. This painful inference would last what seemed a lifetime of mockery until I came to better understand what dyslexia and APD are, having nothing to do with intelligence, and everything to do with how my brain processes the world differently than most other people.

To frame it in simple terms, when I see 2X2=4 written on a blackboard, my dyslexic brain doesn't very easily process the concept, but when I see 2 pennies, next to 2 pennies, then it all makes sense, pun intended. If I hear a phone number and write it down, I usually get all the numbers right, but frequently in the wrong order. Looking up a name and number in a phone book can be a nearly impossible task for me, while just simple reading is a constant brawl between what my eyes see and what my brain doesn't. I can barely spell my own name, but know when a word on paper is spelled wrong. How messed up is that? Dyslexia, a learning disorder, affects your ability to read, spell, write, and speak. Not much left, except perhaps sex, but at my age, even that's a challenge these days! Dyslexia stems from differences in parts of the brain that process language. Imaging scans of people with dyslexia show that areas of the brain that should be engaged when a person reads aren't working properly.

In stark contrast to the pitfalls of dyslexia, there are a number of interesting and beneficial attributes to the so-called disorder. According to research, people with dyslexia often see things more holistically, have the ability to see how things connect to form complex systems, demonstrate better skills at manipulating 3D objects in their mind (I built our house, not from drawings on paper, but drawings in my head), tend to think in pictures rather than words, have better peripheral vision than most, are highly creative, and are well-known for having sudden leaps of insight that solve problems with an unorthodox approach. So perhaps dyslexia is not a disorder at all, but just one of many unconventional brain designs that evolved in nature to help an individual better interpret the complex world we live in.

To quote Matthew H. Schneps, an astrophysicist with dyslexia, "Whatever the mechanism, one thing is clear: dyslexia is associated with differences in visual abilities, and these differences can be an advantage in many circumstances, such as those that occur in science, technology, engineering, and mathematics. In physics, we know that an engine is capable of productive work only when there are differences in temperature, hot versus cold. It's only when everything is all the same that nothing productive can get done. Neurological differences similarly drive the engine of society, to create the contrasts between hot and cold that lead to productive work. Impairments in one area can lead to advantages in others, and it is these differences that drive progress in many fields, including science and math."

Digging into our ancestral past, I wonder if there was some evolutionary advantage to dyslexia? Perhaps the sentinels that slept at the edge of a village, keeping everybody safe, were in fact dyslexic. A person with better peripheral vision might see danger sooner, giving that "**flight response**" a split-second edge over a sable-toothed tiger sinking its 6-inch fangs into the back of your neck. Or the ability to quickly solve a complex problem might give one hunter an advantage over another. Just food for thought!

Although I spent a fair part of my life thinking I was missing a few gears upstairs, it wasn't until my wife Judy, an educator, administered a non-verbal IQ test called the Ravens Standard Progressive Matrices (RSPM) to me that I understood the gears were still there, but turning differently than most. With the task of administering this test to over 400 third graders in the school system ahead of her, Judy asked if I would be her "guinea pig" and I reluctantly agreed. I mean, did I really want to be told yet again how dumb I was? But being the supportive and dutiful husband that I am, I relented. Just get it over with, I thought to myself. But much to my surprise, when the results came back, I had scored significantly above average! Imagine my surprise!

Judy explained that the Ravens is a nonverbal test that removes cultural bias and language barriers present in most other tests that measure general intelligence and abstract reasoning like the WISC-R. With no written words, the test is comprised of progressively more difficult pictures and patterns. The Ravens can be administered to a wide variety of ages from age 5 to the elderly with scores tied to percentiles that account for age differences. So, my reading struggles due to undiagnosed dyslexia during my early school experiences prevented me from scoring well on a traditional IQ test, but the Ravens removed those barriers, allowing me to recognize my full potential for the first time in my life.

The fact that I have authored scientific books, journal papers and given university lectures is a testimony that even a screwed-up brain like my own can overcome reading, spelling, speech and writing handicaps with enough passion and fortitude, not to mention a supportive wife. Interestingly, I only recently discovered that folks such as, Stephen Hawking, Leonardo da Vinci, and Pablo Picasso, were also dyslexic, so I am at least in good company.

### The Anxiety of College

Leaving the Dyslexic calamity stuff behind (I wish), the fear of returning to college failed to derail my newfound aspirations of becoming a biologist, so back to school I went. Up until college, my life was full of rich experiences and learning acquired from a variety of jobs and extraordinary adventures. Ironically, it wasn't until I got to college that my education slowed to a snail's pace and ceased to make a whole lot of sense. From my point of view, college was more about "quantity, not quality of education". For me, sitting hour after hour, week after week, month after month, listening to pretentious professors suffering from mandible-diarrhea, filling notebooks with endless details, just to later regurgitate it all for some arbitrary grade was an exercise in futility. This did little to motivate my creative juices. As expected, I did rather poorly during my time at the University of Cincinnati, but to my surprise, exceedingly well at Hocking College, a result of "learning by doing" as opposed to "learning by reading". What a concept! This, however, is exactly how my backward dyslexic brain learns anything new.

While I considered my education at the University of Cincinnati to be a complete waste of time and money, Hocking College was vastly different. We didn't just read about trapping grouse, we actually went to the woods and trapped grouse. But Hocking College had its shortcomings as well. The primary focus was on animals that carried jumbo racks or protracted beards. Don't get me wrong, I have nothing against hunting wildlife for food. Trophy hunting is a different story, at least in my mind. This, however, was the mindset during the 1980s, namely teaching students to manage wildlife for maximum results (antler size) and/or numbers. But I was more interested in non-game species. In other words, critters that form the actual framework of properly functioning ecosystems, which in turn sustain larger life forms such as deer and turkey.

## US Forest Service and Beyond

Moving forward, it wasn't until serving as a biologist for the US Forest Service in Red River Gorge, that I would have an opportunity to work with nongame species such as the Virginia big-eared bat and White-haired goldenrod. Of all my 9 to 5 kinds of jobs, the US Forest Service was by far the best of the best! I cherish the many years of employment that felt more like play than work. It was truly a life of passion, learning, adventure, and friendships.

My greatest biota enlightenment, however, did not come from college nor the Forest Service, but from biologist, naturalist, and mentor, John MacGregor. John, who could write a best seller on "dry ice" firmly anchored my inquisitive nature on nature's many echelons. From this point forward, nearly all of my knowledge of flora and fauna has been self-taught. College, the Forest Service, and meeting John MacGregor were only stepping stones for better things to come. As it turned out, I ended up quitting the US Forest Service at the ripe young age of 47 to illustrate a book. I know, why would anybody in their right mind quit such a great job to draw pictures? That's exactly what my dad said!

My good friend John MacGregor was beginning a book project (The Reptiles and Amphibians of Kentucky) and was given 2 unpaid years off from the Forest Service to complete the work. John asked me if I would be interested in doing scientific illustrations. I, of course, agreed to be a part of this worthwhile project and approached my supervisor at the Forest Service with a request for 3 unpaid months off to do the drawings. He unenthusiastically agreed.

But after 6 weeks, he reneged on the promise. It basically boiled down to, "*Dan you need to decide what's more important, your job or the book.*" Two weeks later, after some soul-searching and lengthy discussions with Judy, I turned in my resignation with this to say, "*My commitment to John MacGregor and the University of Kentucky Press is more important than working for an agency that breaks promises and lacks the foresight to see the bigger picture and the benefits of inter-agency cooperation,*"—an agency unable to "*rethink and relearn*".

A second, and perhaps more concerning reason I left the Forest Service was Kentucky Heartwood, (an environmental group) and their so-called ecological objectivity and ethics. **It's a story that is seldom told**. While I would miss the fulfilling wildlife management portions of my job, I would lose no sleep over the ever-increasing "bureaucratic bull-shit" brought to my desk by Kentucky Heartwood's endless lawsuits. Continuing to work for the Forest Service would mean spending less time in the field and more time chained to my desk—eventually becoming my worst nightmare (besides flying)—a pencil-neck!

The usual way that environmental groups disrupt agencies like the US Forest Service is to misuse the "Endangered Species Act" and to convince PhD experts to side with them, no matter what the cost. I have several well-respected colleagues with PhDs that outright lied about biological issues, just to bury the Forest Service in paperwork, throwing out their ethics and science for personal sentiment.

In Heartwoods' inexperienced attempts to fix a broken ecosystem, they made uneducated and unreasonable demands to stop any and all FS projects, from building cave gates to protect endangered bats, to digging ponds for frogs and salamanders. "Kentucky Heartrot" (my fond name for them) cared about endangered species such as bats only to the point they could use them to browbeat the Forest Service into submission. But like a used condom, had no use for them following a successful lawsuit!

The mindset of "Kentucky Heartwood" at the time was, *habitat doesn't need our intervention or manipulation to survive*. While this might have been true before *Homo sapiens* showed up on the scene, in this day and age, it is an idea that is wholly unrealistic. Why? Because we have fucked things up so badly, that without human intervention, many species will simply disappear.

I was involved with other lawsuits including one in West Virginia over, believe it or not, a land snail, *Triodopsis platysayoides* or Flat-spired Threetooth. Friends of Blackwater (another one of those environmental groups) sued a logging company over the snail even though the logging company was following all the guidelines set forth by the US Fish and Wildlife Service. "Friends of Foulwater" won the lawsuit anyway, as a result of a colleague of mine making "unsubstantiated scientific statements" which ultimately drove the timber company out of business, along with its many employees.

Blackwater also accused me of being a shill for the timber company for simply presenting the ecological requirements of the snail, which didn't align with their agenda. Based on my research, (the most extensive at the time), it was "rock talus" not "trees" that turned out to be the most critical factor in the snail's survival. But for Blackwater and my dishonest colleague, the scientific facts and my reputation mattered little. And we wonder why people are opposed to the endangered Species Act and environmental groups such as Heartrot and Foulwater.

Not surprisingly, when "Blackwater" was approached by West Virginia DNR to use some of the lawsuit funds to further study the snail, they promptly declined, once again illustrating that critters like bats and snails are important to these groups only to the point they can be exploited, after which they are discarded like rubbish.

On another occasion, while netting for the endangered Indiana Bat in North Carolina, I was approached by an angry logger who said he would kill every bat I caught in the net. He was rightfully upset because the rare bat, if found, could potentially shut down his logging operation, affecting his income and ability to feed his family.

I asked him, "How do you like your bat, stir-fried or boiled? After all, if you are going to kill them we should go ahead and eat them." He laughed aloud and things began to calm down. As we talked a bit, our conversation became more like two good old boys. He then asked me "What's more important, people's jobs or bats?"

It was a great question, and at first, I had to ponder. I eventually responded, “Both are equally important. After all, my job is to protect the bats which eat the bugs, that harm the trees, you cut down, to build our houses and make the toilet paper to wipe our asses." When asked the same question, most politicians respond by saying "JOBS, of course," never seeing the bigger picture. Unfortunately, this kind of lopsided thinking has set the tone that has spread across the nation like a California wildfire.

One more gripe of mine is the "environmental types" that stand on hilltops in Eastern Kentucky and West Virginia, demonstrating against mountain top removal, but afterward, go home, turn on lights, and pop a cold one from their refrigerator. Don't get me wrong, I come from a short line of tree-huggers, so I get it! But while I agree that this type of mineral extraction has done serious ecosystem damage, I also realize that moving toward "questionable renewable energy" will take time and still require a shit load of fossil fuels to get us there.

Heartwood, through their constant harassment followed by lawsuits, changed forever, the essence of the US Forest Service, some of which was good, much of it not. But in their attempt to swing the pendulum all the way to one side—theirs—Heartwood lost sight of the more important goal; a better balance within the agency. They were so hyper-focused on the cause—they lost sight of the truth!

I would like to think that people such as myself and John MacGregor (who never returned to Forest Service) brought a balance to the agency, which was lost as a result of Heartwood pushing the Forest Service too hard. Consequently, it was groups like Heartwood that "**drove the passion out of the Forest Service**", leaving old farts that think the only good tree is a stump. In the end, my job with the Forest Service was beginning to feel more like work, not passion and therefore, when the ultimatum was issued, it was time to seek out new adventures, to boldly go where I had not gone before!

I finished the biological illustrations on time and moved on to becoming a private biological consultant for the National Park Service, US Fish and Wildlife Service, and Nature Conservancy. My newly-discovered adventures would also include spending 7 years in a Belizean rainforest managing a biological field station BFREE (Belize Foundation for Research and Environmental Education) with my wife Judy, researching land snails in the Great Smoky Mountains National Park as part of the ATBI (Discover Life in America Initiative), a National Geographic Expedition in Belize, research in the Amazon Jungle, and writing a number of natural history books and research papers.

In conclusion, while most biologists are drawn toward the more flamboyant wildlife, I seek out those less charismatic and much-maligned creatures; the underdogs of the world. Why? because they are a lot like me, but without a voice. So this is my calling, my passion, my endowment to give a platform to the ones that do not speak, to sing their songs, and tell their story. For they are the vastly misunderstood and undervalued creatures that form the very framework of our living planet. And this is my convoluted journey that eventually brought me to Belize and beyond.

### Nuptials of Grief and Joy

A discussion of my life would seem incomplete without a brief mention of my two marriages and one spawn, Angie. Around the time at Hocking College and before my job on the Sam Huston National Forest in Texas, I met my first wife Grace. We were in Catholic grade school together, but hadn't seen each other since those awful days under the watchful eyes of sex-deprived nuns. It was lust at first sight, as it is for all, and so grew the romance. My plan at the time was to finish collage, return to Kentucky, build a small house, and live a simple, back-to-nature kind of life. At the time, she thought this sounded great and was totally on board, or so she thought. But it was a bittersweet 5-year marriage that did not end well. The bitter for me was being married to a person that had behavioral symptoms analogous to OCPD (obsessive-compulsive-personally-disorder) and narcissism that I was clearly ill-equipped to deal with. I know this sounds cliché, but after we were tethered, Grace went from unruly and fun to serious as a newborn puppy looking for that first drink of mother's milk. It was as if somebody or something (Mormonism?) turned on the "serious-switch", robbing "**who she was**", making her "**who she wasn't**".

There's something to be said about living together (which we did not) before a hitching. It gives you time to get past that dopamine-induced, lust-phase. Nevertheless, if God had been smarter, he would have made us all **hermaphrodites**, doing away with gender identity, inequality issues, and the endless problems associated with male and female relationships.

In all fairness though, the failed marriage was as much my fault as it was hers. While my animal skills were/are fairly fine-tuned, my people skills have much to be desired. According to my wife Judy and daughter Angie, I often lack sensitivity to the everyday, mundane emotional needs of others, which I would agree with, but when those emotional needs reach critical levels, my availability and compassion are rock solid.

So my lack of listening aptitude and sympathy to Grace's daily emotional needs was probably my biggest failure as a partner. As it turns out, my sensitivity issues appear to be tied to having "Aspies". More on this dreaded syndrome later. A deficiency in religious conviction was no doubt a factor as well.

If truth be told, **SENSITIVITY** is why all men are assholes and all women are bitches, having either too much or too little of it, never really getting the mixture quite right. In any case, Grace and I were doomed from the start and the good times we shared were few and far between.

The worst years of my entire life, however, would be our divorce, when Grace decided to move from Kentucky to Arizona, tearing my 3-year-old daughter from my life—a separation for Angie and I that would seem to last a lifetime. But for me, it was worth all the pain that Grace and I suffered for that singular moment in time when our daughter Angie came forth and took her first breaths of life. Without question, I consider Angie my greatest achievement in this crazy and mixed-up world and I am more than honored to be called her father and friend.

Then along came Judy, Judy, Judy. While wife number one was a mismatch of souls—actually brains, wife number two was leaps and bounds different. Like the disparity between milk and dark chocolate, the latter being nearly perfect, or about as good as it gets! Before I met Judy however, I was a crushed and lonely man of 7 years that had been dating three basic kinds of women: pleasantly plump, athletic-types with over aggressive personalities, thin ditzy blonds that thought shopping at Wal-Mart was a good time, and those infrequent and terrifying encounters with attractive but psychotic females. It was a smorgasbord of women wanting to either hunch-a-bunch or breed for one of those ugly, fat-headed, no-neck babies that everybody thinks are so adorable. Let's be honest, newborn humans are hideous-looking creatures, nowhere near the cuteness of newborn puppies.

I had become somewhat of a curmudgeon and heading for a life of loneliness and destitution, at times, ready to end it all. *Ok, it wasn't quite that bad.* But after losing my daughter to distance and crying myself to sleep every night, I learned to callous my feelings for Angie and in the process, lost a significant part of who I was. Judy would be the catalyst between my daughter, reigniting my feelings and bringing me closer to Angie. It was by far Judy's greatest gift to me, and I am forever in her debt. We have been married 25 years now—wait, Judy says it's been 27 years. I will never understand how and why she remembers this mostly useless information, but after decades of teaching her bird calls, she still can't tell a Tufted Titmouse from a Dickcissel.

I know what you are thinking though; why in God's name could you, or would you ever forget an important date like your anniversary right? It's quite simple really. I prefer buying gifts from thoughtful, unfettered positions over obligatory social constraints. Why? Because the action comes from a more sincere place, not the media-driven "special day" bullshit that drives our everyday buying decisions. And guess what, Judy appreciates the unanticipated and spontaneous thoughts more than the "I'm owed something on this particular day". Even when we were hitched, we did not exchange the gold and diamond, **"LAND-RAPING"** bands of marriage. What a rare woman she is indeed!

Moving on, while Judy and I have had our share of bumps, she has been a guiding light in my life. Most importantly, Judy has helped me reach my fullest potential, giving me the confidence to do things that I would have thought highly unlikely for my dyslexia brain, such as writing books or publishing research papers in scientific journals. But my marriage story doesn't end here and would be incomplete without some conversation about our three spawns.

## Offspring

When we got together, I was thrown into fatherhood, but notwithstanding a unique set of challenges. Judy's two ankle-bitters, Brad and Tyler, would often remind me that I was not their father, and as a result, didn't like me telling them what to do. I would slowly earn their respect, however, by taking them caving, canoeing—you know, kids stuff'. But what really cinched the deal was our first Halloween together. I went all out and bought them each a whole bag of Hersey's candies, which at the time blew them away, or at least in my direction.

I learned quickly that the fastest way to a kid's heart is through their cavity-filled, saccharine-daggers. While it took time for all of us to come together, our slightly quirky but well-blended family is currently tighter than a frog's bung-hole in pond water. Raising her two boys was my honor but I often tell them that while I did all the work their biological father got all the credit. I know this isn't true, of course, and love Brad and Tyler as my own, but let's keep it our secret!

### The Unearthing of Asperger's

It wasn't until I started writing about myself that the universe would yield yet another twisted sense of irony. As if Non-white Vision Impaired, Auditory Processing Disorder (APD), and Dyslexia weren't enough handicaps to saddle a person with, the universe decided to shoulder me with one more disability, Asperger's Syndrome. When I first heard this term decades before, I thought it was a person with an overabundance of ass-boogers or dingleberries! Little did I know at the time I actually had the dreaded condition. True to my diagnosis and nature, I jokingly referred to this ever-so-serious topic in an attempt to come to terms with this unexpected diagnosis.

While my wife has for years accused me of having Tourette's syndrome due to my repetitive, sing-song phrases that make no sense in context and change when I get bored with them, my epiphany regarding my many strange behaviors occurred while watching a YouTube video of climate-activist Greta Thunberg. As she discussed her "disability", Asperger's Syndrome or Autism Spectrum Disorder (ASD), all the symptoms she described resonated with me, especially this particular sentence.

*"Those of us on the Autism Spectrum see things as black and white, are not fond of lying, and don't enjoy the social games most of you are so fond of." Greta Thunberg, TED Talk-Stockholm.*

That's so me! A little further research led me down the rabbit hole of autism and the autism spectrum and to this simplified definition.

*Asperger's syndrome is a condition on the autism spectrum associated generally with higher functioning individuals that may be socially awkward and have an all-absorbing interest in specific topics.*

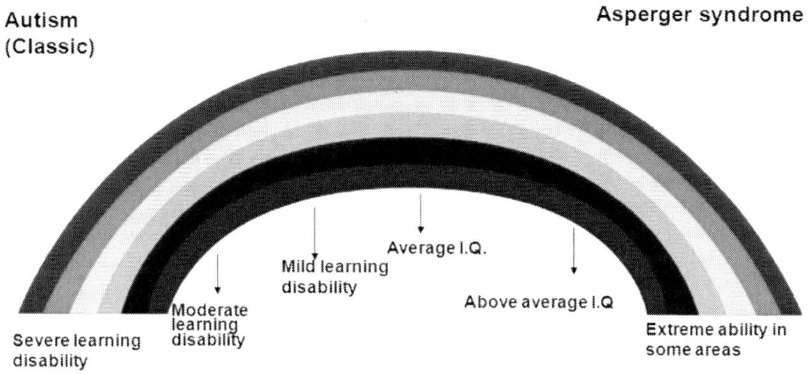


The graphic above helped me map my place on the Autism spectrum. As I learned more about this neurodivergent way of thinking, the more I understood myself and all my weird, unexplainable behaviors. Could I have Asperger's Syndrome on the Autism Spectrum, I wondered?

Throughout my life, I had always just attributed my oddness to just that, being odd, never really fitting the mold of what most consider as normal. To finally be able to put a name to it was both comforting and disconcerting at the same time. Wasn't autism considered to be a negative thing to have? Below are some common characteristics shared by those with Asperger's with my comments.

- **Routine-oriented.** *That's me in a nutshell.*
- **Uncomfortable in situations where the person feels like they do not have control over a situation or place.** *My fear of flying! I have often said, it is not the actual flying but the feeling that I have no control in the back. of a tin can with wings.*
- **Unintentionally rude or uninterested in others.** *My wife often gives me "the look" like the time I saw a person that I had known as a child who was now considerably heavier than I had remembered. I said, "Wow, you gained so much weight, I didn't recognize you." I guess it's ok to say just the opposite though.*
- **Tendency to prefer to be alone over social interaction.** *I am totally content to be at the house for weeks without leaving.*
- **Repetitive behaviors**. *This one is particularly interesting and comical to my wife and immediate family. Over the years, I have adopted*

*repetitive, often nonsensical, almost lyric phrases or sounds that I randomly repeat multiple times throughout the day. These include but are not limited to "Little, little, little...little dog" or my current favorite "Today" (adopted from a silly song about bananas while we were in Costa Rica with the grandchildren). I retain the "songs" until something new hits me. How messed up is that???*

- **Speech Difficulties (especially as a child).** *I didn't talk till I was five years old and had speech therapy as a kid.*
- **Children with Asperger's are often isolated with few friends.** *This certainly was me. Even into high school, my circle of friends was very small (3) and included other often odd and ignored kids.*
- **Poor sleep patterns. Difficulty falling asleep and is easily awakened.** *An ant takes a piss wakes me up at night!*
- **Hyper-focused on subjects of extreme interest often becoming experts in a field.** *This one really hit home with my hyper interest in the study of land snails.*
- **Lack of Empathy or Sensitivity.** *New research suggests what may appear as a lack of empathy is just the opposite, more likely overwhelming empathy that causes the person to bury emotions as a self-protection mechanism. When my daughter was ripped from my life, I developed an emotional callous to survive the awful ordeal.*

In my convoluted journey of self-discovery, I wondered, "Were my *conditions* the result of some evolutionary advantage to earlier humans that was somehow left behind?" For example, in prehistoric times sensitivity was probably the last thing on our minds. After all, we were hunters and gatherers—spending much of our time scouring the land for food—while avoiding being gored by a woolly mammoth or eaten by a cave bear. As we moved from being hunters and gatherers to farmers and animal domesticators, we changed forever our fundamental nature. As our lives became increasingly easier, we had more time to cultivate emotional needs such as autonomy, competence, and relatedness—which in turn amplify one of the most defining human traits, sensitivity.

But it wasn't without its baggage, and today we are a society full of hypersensitive individuals that whimper at the drop of a hat. Coupling relationships tend to bring out the worst imbalances of sensitivity—each individual having either too much or too little of the attribute—making it difficult to get along. So while dyslexia and Asperger's might have been advantageous in the past to survive in a world of hunting or be hunted, today these brain disorders might seem rather out of place.

## The Myth of the Normal Brain: Embracing Neurodiversity

While there is a standard for measurements of weight (kg) or a gold bar in Fort Knox is the "gold" standard, there is no brain in a vat on the shelf of the Smithsonian that represents the normal, standard human brain. How do we decide what is normal or abnormal when it comes to brain function or learning? Significant discrepancies exist in determining the lines between normal neurological behaviors and abnormal. In the past few decades, research has blurred the lines by suggesting that neuro-disorders possess significant strengths as well as weaknesses with those strengths rooted in our evolutionary past.

In my case, my proclivity to systematic, hyper-focused problem-solving as a result of Asperger's might have been highly advantageous for the survival of prehistoric humans. Temple Grandin (a fellow Aspie) speculated, "Some guy with high-functioning Asperger's likely developed the first stone spear; it wasn't developed by the social ones yakking around the campfire." Upon reflection of Ms. Grandin's statement, I had yet another epiphany about my unconventional self which further supported the idea that this "ass-boogers" thing has been more of an advantage to me than I would have ever imagined. My Auditory Processing Disorder, Dyslexia, Autistic (APDDA) brain has actually allowed me to "invent" devices and methods to simplify procedures or more efficient ways of doing things.

During one particularly boring Forest Service biologist meeting, my mind wandered afar (as it so often does) from the drone of the speaker to a problem regarding bat house designs that failed to attract bats. Instead of thinking like a human to solve the problem, I began thinking like a bat and sketched in my mind a new bat house concept (mimicking a dead tree with exfoliating bark) that would become the Rocket Box.

On another occasion, the desire to help my struggling daughter resulted in what I consider my most important creation (besides her), the **Callie-Can** (page 206). When my third grandchild was diagnosed with Cystic Fibrosis (CF) at 8 days old, it sent the entire family into a tailspin of emotions and time-consuming procedures to keep her thriving that would quickly overwhelm my daughter. In order to keep Callie's lungs from becoming clogged of sticky mucus, my daughter was required to perform chest beatings on each of four quadrants on her newborn's chest, both front and back, for a minimum of 5 minutes on each spot, a procedure that would take 40-45 minutes three times a day.

Not only time-consuming, the manual procedure was physically draining to the administering person. Our son-in-law assisted when he was not at work but much of this procedure fell on my daughter's shoulders to perform and she was quickly feeling overwhelmed. "I wish there was some way to speed up this procedure," she lamented to me. After some brainstorming with her, I sketched a crude design in my head and headed off to the local hardware store. A few hours later, the first prototype of the Callie-Can was born. The device would hit all four quadrants at once, thereby cutting the time in half, from 40 minutes to 20 minutes! I will never forget the feelings of sheer joy and accomplishment I felt when she used it for the very first time and it worked, seeing tears of joy stream down her face. So the realization that "Asperger's Syndrome" may have been the reason I was able to invent something to help my daughter in her struggle is totally worth some of the negative connotations associated with autism. Later, Callie had a feeding port added to her stomach because she was not gaining any weight. Angie and Colby had to push liquid food through a narrow tube via a syringe which was nearly impossible by hand. My awkward gears began turning again and the Juiceinator was born (page 206).

Another idea that came to me while walking the Pacific Crest Trail in 1980 was having a water bottle, connected to a hose that could be accessed while walking, eliminating the need to take off my heavy backpack every time I wanted a drink. Nine years later the CamelBak was invented when, Michael Eidson, an EMT repurposed an IV bag as a way to carry extra water on the Hotter 'N Hell 100-mile bike race outside Wichita Falls, Texas. While ideas come easy to me, taking them to the next step is where I struggle the most!

Harvey Blume, one of the first to refer to autism and other mental disorders as "**neurodiversity**" noted, *"Neurodiversity may be every bit as crucial for the human race as biodiversity is for life in general. Who can say what form of wiring will prove best at any given moment? How absurd it would be to label a calla lily as having "petal deficit disorder" or to diagnose a person from Holland as suffering from "altitude deprivation syndrome. There is no normal flower or culture. Similarly, we ought to accept the fact that there is no normal brain or mind."*

Upon further research, I learned that many of the most successful people throughout history were either confirmed or thought to have been on the Autism Spectrum. It includes folks like Bill Gates, Albert Einstein, Steve Jobs, Dan Aykroyd, Temple Grandin, Anthony Hopkins, Carl Sagan, and Wolfgang Mozart.

Although I have not achieved anywhere near the level of notoriety these individuals have, when I reflect back on my work with land snails, my hyper-focus for these underdogs of the animal kingdom have perhaps contributed more than I realized. For example, the intense study of land snails led me to write 7 books on the subject, clearly putting me on the obsessive side of normal! But according to the malacological community, my work in land snails has been seen as a milestone.

Another frustration of mine caused me to completely rethink scientific keys. Many books on animals fail to anchor the beginning naturalist because dichotomous keys are typically written by people who don't need them (the experts), for people who can't use them (the novice). This is where science has failed miserably, creating a disconnect between itself and the layperson—and it's why most people have, in the past, cared so little about non-charismatic animals such as land snails. I attempted to bridge this gap by using my backwards thinking brain to create books that are both useful to the beginner and expert alike. I began with the simple notion that I knew nothing about snails, therefore removing much of the knowledge-bias, in turn allowing me to rethink and simplify the keys. Going against scientific standards however is notwithstanding criticism from colleagues—even though my nonconformist snail keys have been used by 12 year olds successfully.

While some of the behaviors associated with Asperger's caused me to suffer as a child, there are many positive attributes which I now embrace. Some of the optimistic values that have caused me to celebrate not malign this diagnosis are as follows.

<u>Positive Attributes of People with Asperger's:</u>

- are able to easily forgive others
- are enthusiastic and have a propensity for obsessive research, thus developing a broad and deep base of knowledge in subjects of interest
- are intelligent and talented
- are more likely than those of the general population to pursue a university education
- are not inclined to lie
- are not motivated by an intense social drive to spend time with whoever happens to be available
- are persistent, and when they set their minds to something or make a promise, they can usually be trusted to follow through.
- are very accepting of the quirks and idiosyncrasies of others
- bring a highly original perspective to problem-solving
- can recall fine details that others miss
- have an acute sensitivity that supports creative talents
- have exceptional memories
- have extreme endurance

- have values that aren't shaped by religious, financial, social, or political influences
- stick to their positions, even in the face of intense social pressure *(why I don't smoke, drink, or do drugs)*
- tend to prefer individual sports to team sports, as there are no social demands and they can exercise complete control over the activity

Negative early childhood educational experiences compounded by a lack of understanding about Asperger's left me feeling ostracized and incapable. Yet there are those life-changing moments when a not-so-ordinary teacher, Mrs. Rath, recognized not my weaknesses, but my strengths. It was during a 4th grade social studies project for school. My chosen venture was an elaborate scale model of the Galapagos Islands I built on a piece of plywood complete with stilted-huts and a volcano. But it didn't end there. My detail-oriented brain insisted that in order for the project to be complete, I needed live turtles to represent the extraordinary giant land tortoises that lived on the islands. A supportive and loving mom scoured the city and eventually found 2 baby Pond Sliders. The project was complete and Mrs. Rath recognized it as being the most creative in the class.

While much of my educational memories are scarred by a plethora of teachers who did not understand my neurodivergency, the encouragement and recognition of my abilities by that special educator, helped me believe in myself. This, and other safe islands like my parents and nature, were lifelines that kept me just barely above water, until later years when I began to better understand my so-called disabilities. But when it came to writing books, Judy gave me that last helping hand to climb to a place of self-confidence.

So maybe, this "ass-boogers" thing wasn't so bad after all. Over the years, my Aspies has no doubt helped me preserve my individuality—even as others are surrendering theirs to the hordes! And while I am no Mozart, sharing similar qualities to some of the remarkable individuals such as Temple Grandin softens the blow of yet another condition that has held me hostage for most of my life. Nevertheless, it is exhilarating to still be learning profound things about myself at age 67! And when my wife says I'm being insensitive or calls me an asshole, I now correct her by saying "NO, "I'm being an **Aspie**"

The Callie Can

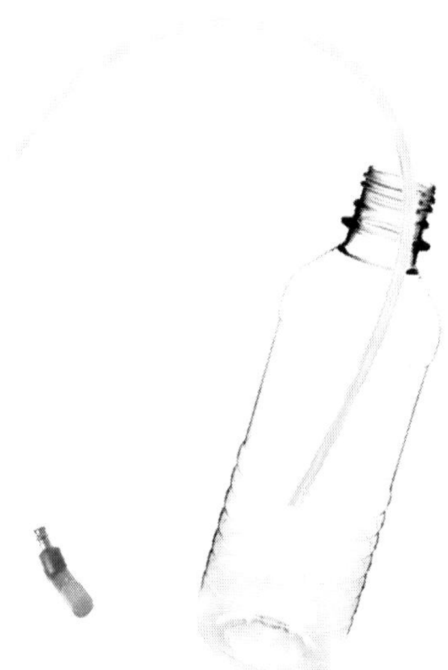

My Backpacker Water bottle

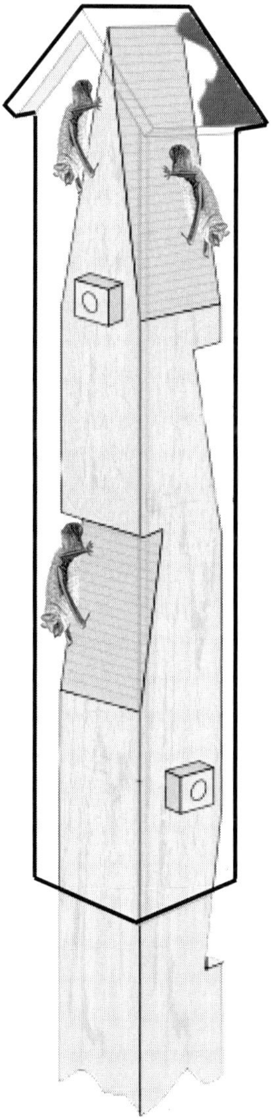

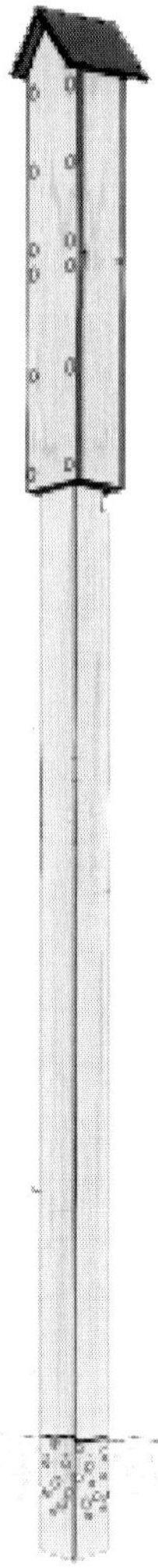

The Rocket Bat House

The Callie Juicinator

# My World Views in Short

## *Dannerisms: Personal Philosophies*

*If it feels like work, I stop doing it.*

*Dying is a privilege because we got to live.*

*If work is what we dread and play is what we enjoy,*
*why not choose a job of enjoyment over a job of trepidation?*

*I don't make my life about money, but about passion.*

*The only person that keeps you from learning to paint, or play the guitar,*
*writing a book, or going to Belize with your wife to manage*
*a Biological Field Station for 7 years is you!*

*Everybody lies but not everybody is a liar.*

*Be happy with who you are, not who you are not.*

*Let passion and wonder guide your life's journey.*

*We need to stop burying our dead and start planting our dead.*

*We are a species hopelessly addicted to bullshit.*

*Grab life by the balls or be the balls that are grabbed.*

*The most liberating and honest day in your life*
*is when you stop believing in Gods.*

*We all break the laws, but that doesn't make us all lawless.*

*Stop doing the right thing for the wrong reason, a milk-bone.*

*Many believe if there's no God there is, no purpose.*
*I give my life purpose not a fairy tale.*

*Churches and graveyards would make better*
*homeless shelters and green spaces.*

*Lust is blind, love is clarity.*

*Aphrodisiacs are for those men who lack the*
*balls to accept the inevitable.*

*A closed mind is a terrible thing to waste.*

*We are not a species that has lost our minds,*
*we are a species lost in our minds.*

*Relationships are like a box of chocolates, the outside*
*always looks delectable, while the*
*inside is an uncertainty.*

*All life on Earth is a result of an astronomical number of chance encounters. Consequently, life is an extraordinary privilege. But life with consciousness is, without doubt, our single greatest endowment, taking human existence to unparalleled levels of understanding.*

*In our fleeting moments in time, may you find solace and unassuming kinship with Earth's infinite wonders.*

*I don't consider myself an owner of land*
*but rather a steward of land.*

*Religions don't want you to be who you are, but who you're not.*

*I will celebrate your success, but not wallow in your pain.*

*Telling truth is always better than hiding truth.*

*You can't change genetics, but you can certainly give them hell.*

*We need to teach our kids how to think, not what to think.*

*The universe doesn't give my life meaning, that's for me to decide;*
*the universe doesn't tell me how to live, only a place to live,*
*the universe is indifferent to my successes,*
*my failures and my passing.*

*CEOs, like politicians, should be voted upon by the people.*

*Normalcy is like a cancer that slowly eats away at individuality.*

*Our greatest nemesis will always be our thinking brains.*

## Others Ideas I Embrace

*Life is either an adventure or it's nothing.*

*Let go or be dragged.*

*Don't get attached to outcomes which you have little or no control over.*

*In order to solve a problem, you must first acknowledge there is one.*

***Shit just happens!***

# In Belize

*A song written by our good friend*
*Ann MacFie from Kentucky*

The limes are orange, the oranges are green
You can eat beans and rice or eat rice and beans
In Belize. Wanta go back to Belize.

The sand is so white, the water so blue,
The fish on the bottom can look up at you
In Belize. Wanta go back to Belize.

The beach and the islands, the reef and the sea
The savannah, the mountains - it's so fine to be
Coming home to a world like the one we all came from before.
The toucan, the tapir, the shy manatee,
The macaw, and the margay - all living free
In their peaceful, green home, where the great harpy eagle can soar Over paradise

Every house is painted so gay.
It might be just one room, but those bright colors say,
"Belize!" Wanta go back to Belize.

Everyone's outside and going somewhere
On the rivers and roads, back and forth, here and there
In Belize, wanta go back to Belize.

Red, black and white folks, they get along.
It's a world right out of a Bob Marley song
That blares from a boom box outside the taxi man's stand.
People down there don't have a lot,
But they know how to be happy with what they've got
And they know what they got when it comes to their beautiful land -They got a paradise

The moon rises up and it shines down so bright
It's all silver and shadow, the jungle at night
In Belize. Wanta go back to Belize.

Good night, sleep tight,
Don't let the kissing bugs bite.
Tuck in your net and blow out the light
And listen to the howlers sing -- you get to loving it!

The limes are orange, the oranges are green,
It's the craziest place that you ever have seen
In Belize. Wanta go back to Belize.
Kissed by Caribbean breeze
Washed by the gentle, blue seas
Where the nightwalkers crash through the trees
No di talk politic, please
Everyone smile and say "cheese,"
So they'll know you went to Belize.
Belize, ah, Belize.

# Monkey Poo Cookies

Ingredients

2 1/2 c. whole wheat flour
1 1/4 tsp. baking powder
1/2 tsp. baking soda
1/4 tsp. salt

1/4 c. olive oil
2/3 c. natural peanut butter
8 T. butter
1 1/2 c. brown sugar
2 eggs
2 1/2 tsp. vanilla
2-3 T. molasses

1 c. chopped nuts
1 12oz bag of semi-sweet chocolate chips

11/4 cup raw non-sweeten cacao power
(the secret ingredient)

Instructions:
Preheat oven to 350° F. Grease a cookie sheet.

Mix dry ingredient in medium bowl. Mix wet ingredients in a large bowl. Gradually mix dry ingredients into wet ingredients. Add nuts and chocolate chips.

Mix well.  Using a teaspoon, scoop dough and roll into balls. Place on greased cookie sheet.  Smash balls with a fork. Bake for 8 minutes. Allow to cool for 5 minutes and then ENJOY!

# The Five Native Cats of Belize

Jaguarundi

Puma

## About The Author

Dan Dourson is a conservation biologist/naturalist/illustrator/lecturer who has dedicated much of his life to the preservation, conservation, and understanding of the planet's more obscure species. A wildlife biologist with US Forest Service for nearly 20 years, he worked to safeguard a myriad of species including snails, mussels, bats, reptiles, amphibians, and other creatures often ignored or misunderstood. Dan has conducted biological inventories for the Nature Conservancy, US Fish and Wildlife Service, Kentucky State Parks, Kentucky Department of Fish and Wildlife Resources, West Virginia DNR and worked for more than a decade studying land snails in the Great Smoky Mountains National Park, USA as part of the ATBI, Discover Life in America Project.

Dan has been studying land snails in the eastern United States, Central America, and in the Amazon Jungle of northwestern Peru for over 25 years. In 2016, he was a co-investigator, leading a National Geographic Expedition into the Maya Mountains of Belize, studying a bottom-up relationship between land snails and harpy eagles. Dan and his wife, Judy, spent seven years in Belize where they worked with Belize Foundation for Research and Environmental Education (BFREE). Dan was the resident biologist, assisting with numerous research projects and helping his wife manage the remote biological field station. He continues to work on land snail inventories in eastern North America and abroad. Dan currently lives with his wife, Judy, in the Red River Gorge area of eastern Kentucky in a tiny off-grid house. His allegiance with the natural world is clearly reflected through his writing and simple lifestyle.

**Other Books Written by the Author:**

- *Land Snails of Kentucky and their Ecological Communities*
- *Land Snails of the Great Smoky Mountains and Southern Appalachian Mountains*
- *Land Snails of West Virginia*
- *A Natural History of the Bladen Nature Reserve and its Gastropods, Belize*
- *Biodiversity of the Maya Mountains, Belize Central America*
- *Land Snails of Belize*
- *Wildflowers of Red River Gorge and Greater Red River Basin*
- *Wild Yet Tasty*
- *Reptiles and Amphibians of Red River Gorge*

**Books illustrated by the Author:**

- *How Snakes Work*
- *Wildflowers and Ferns Of Kentucky*
- *Ginseng Dreams*
- *Reptiles and Amphibians of Kentucky (not yet in print)*

Made in United States
North Haven, CT
04 November 2026

21772148R00127